Principles of Plant
Infection

Principles of Plant Infection

J. E. VAN DER PLANK

Plant Protection Research Institute
Department of Agricultural Technical Services
Pretoria, South Africa

ACADEMIC PRESS New York San Francisco London 1975

A Subsidiary of Harcourt Brace Jovanovich, Publishers

ACADEMIC PRESS, INC.
111 Fifth Avenue, New York, New York 10003

United Kingdom Edition published by
ACADEMIC PRESS, INC. (LONDON) LTD.
24/28 Oval Road, London NW1

Library of Congress Cataloging in Publication Data

Van der Plank, J E
    Principles of plant infection.

    Bibliography: p.
    1.   Plant diseases.   I.   Title.
SB731.V28   632   74-10209
ISBN 0-12-711460-2

Contents

PREFACE   ix

Chapter 1. THE RELATION BETWEEN THE AMOUNT OF
            INOCULUM AND THE AMOUNT OF DISEASE IT
            PRODUCES

1.1 The Question   1
1.2 The Simplest Relation: Disease Is Proportional to Inoculum   1
1.3 The Commonest Relation: The Ratio of Disease to Inoculum De-
    creases as the Amount of Inoculum Increases   2
1.4 The Effect of Antagonistic Interaction between Spores: As Inoculum
    Increases Disease Increases to a Maximum and Then Decreases   3
1.5 The Effect of Synergistic Interaction between Spores: as Inoculum
    Increases It Becomes Increasingly Effective   4
1.6 An Undemonstrated Relation: A Numerical Threshold of Infection   6
1.7 An Impossible Relation: Infection without Inoculum. Evidence against
    Recurrent Heterogenesis of Viruses   8
1.8 A Summary. The General Problem: What Chapters 1 and 2 Are
    About   10
1.9 Multiple Infection of Plants: The Problem of Assessing Data about
    the Percentage of Diseased Plants   12
1.10 The Theory of a Numerical Threshold of Infection. Evidence against
     It. Confusion with a Dilution End Point   14
1.11 The Experimental Law of the Origin. The Nature of Experimental
     Law   20
1.12 Curves Depicting Synergism between Randomly Dispersed Spores or
     Infection Sites. Relations near the Origin   22
1.13 Infection by Bacteria   25
1.14 Disputed Evidence for Obligate Synergism between Bacteria in
     Incompatible Host–Pathogen Combinations   27
1.15 The Complex Virus Story. Tobacco Mosaic Virus. Infection without
     Obligate Synergism   29
1.16 Sour Cherry Ringspot, Prune Dwarf, Cowpea Mosaic, Alfalfa
     Mosaic, and Tobacco Rattle Viruses. Obligate Synergism between
     Qualitatively Different Particles of the Same Virus   32
1.17 Incomplete Virus Particles That Infect without Obligate Synergism.
     Potato Spindle Tuber Virus. Satellite Viruses   36

1.18 Restrictions Imposed on the Pathogen by Obligate Synergism. Vector
     Transmission. Possible Benefits from Obligate Synergism   37
1.19 Disease/Inoculum Curves in Relation to Transmission by Vectors   40
1.20 Infectious Entities   42
1.21 A Basic Principle of Infection: The Probabilistic Viewpoint   44
1.22 Disease/Inoculum Relations near the Origin: Independent Action   45

Chapter 2. MORE ABOUT DISEASE/INOCULUM CURVES

2.1 The Scope of the Chapter   48
2.2 The Two Paths to Susceptibility and Their Relation to Curves *A* and
    B of Fig. 1.7. A Hitherto Neglected Factor in Epidemiology   48
2.3 The Relevance of the Two-Path Concept of Susceptibility to Models
    in Epidemiology   57
2.4 Curve *C* of Fig. 1.7. The Antagonistic Interaction of Spores   59
2.5 Curve *D* of Fig. 1.7. The Synergistic Interaction between Spores 60
2.6 A Short Summary of the Chapter So Far   62
2.7 Disease/Inoculum Curves in Root Disease   63
2.8 Tests for Independent Action of Spores   67
2.9 The Use of Logarithms in Disease/Inoculum Studies   68
2.10 Parameters in Disease/Inoculum Relations   70

Chapter 3. EFFECT ON DISEASE OF VARIABLE, LIMITING
            FACTORS OTHER THAN INOCULUM

3.1 The Scope of the Chapter   74
3.2 The Importance of Variation   75
3.3 Multiple Regression Analysis of Variables   77
3.4 The Relative Importance of Inoculum and Other Factors   80
3.5 Temperature and Moisture   80
3.6 The Interaction of Factors Other than the Amount of Inoculum.
    The Interaction of the Amount of Inoculum with Other Factors   83
3.7 Inoculum. Inoculum's Potential. Inoculum Potential   84

Chapter 4. EPIDEMICS. THE TIME DIMENSION

4.1 Time as a Factor and a Dimension: How the Chapters Fit Together   88
4.2 The Basic Infection Rate *R*. The Periods of Latency and Infectious-
    ness. Removals   91
4.3 The Apparent Infection Rate *r*. The Incubation Period   92
4.4 The Two Infection Rates Contrasted and Compared. Historic
    (Memory) Factors   93
4.5 Historic Factors That Increase the Variance of *r*. Cryptic Effects of
    Wavelike Variations. A Failing of Multiple Regression Analysis
    and a Simple Precaution   95
4.6 Historic Factors That Reduce the Variance of *r*. The Principle of
    Continuity   98
4.7 Simulation of Epidemics. Epidem and Epimay. Some Conclusions.   100
4.8 Equations for Infection Rates. Misuse of the Word Logistic   104

4.9 Continuous and Discontinuous Infection. A Simple Rule for Interchanged Estimates 105
4.10 The Period of Latency. Some Objections Considered 107

Chapter 5. WHEN TIME IS UNIMPORTANT. ENDEMIC DISEASE

5.1 Some Definitions. Endemic Disease of Perennial Tissues. Obligate Parasites Used for Illustration 111
5.2 Epidemic and Endemic Disease as a Continuum 113
5.3 An Equation for Timeless Disease 114
5.4 Loss of Infectiousness. Endemic Disease Likely to Be Underestimated. A Steady State Impossible with Obligate Parasites. Anti-Epidemics. Latent Viruses an Exception 116
5.5 Endemic Disease and Native Disease. Sporadic Epidemics of Disease Usually Endemic 119
5.6 The Implications of Endemicity. Horizontal Resistance in the Host Plants. Three Propositions 120
5.7 Adaptation in the Pathogen to Endemic Disease 122
5.8 The Rarity of Harmful Virus Infections in Forest Trees 124
5.9 The Constant Absence of Disease 127
5.10 Summary: Epidemic Disease, Endemic Disease, and No Disease. Host Population Immunity and Resistance Distinct from Individual Host Plant Immunity and Resistance 129

Chapter 6. THE SPREAD OF DISEASE. TIME AND DISTANCE AS DIMENSIONS

6.1 The Spread of Disease and the Migration of Pathogens 131
6.2 Spread a Feature of Epidemic Disease 132
6.3 The Rate of Multiplication (the Infection Rate) and the Rate of Spread of Disease. Dispersal of Pathogens 132
6.4 Spread and Dispersal. Effective Dispersal 133
6.5 Foci of Disease 134
6.6 Varying Gradients of Dispersal. Migration and Colonization 135
6.7 The Multiplication and Spread of *Phytophthora infestans* Seen as Concurrent Processes 138
6.8 A Correlation between the Infection Rate and the Rate of Spread of Disease 140

Chapter 7. GENETICS OF HOST–PATHOGEN RELATIONS

7.1 Introduction 143
7.2 The Fundamental Classification of Resistance. Differential and Uniform Interactions between Host and Pathogen. Vertical and Horizontal Resistance 143
7.3 Flor's Gene-for-Gene Hypothesis 145
7.4 A Second Gene-for-Gene Hypothesis. Gene Identity and Quality 147
7.5 Potato Blight: The Resistance Genes R_1 and R_4, and the Frequency of Virulence on Them in Population of *Phytophthora infestans* 148
7.6 Wheat Stem Rust: The Resistance Genes Sr_5, Sr_6, and Sr_{9d}, and the Frequency of Virulence on Them in Populations or *Puccinia graminis tritici* in Canada 152

7.7 A Theory of Leaky Mutants and Missense Mutation   157
7.8 An Essential Postulate. Some Light on Dominance and Allelism   158
7.9 The Commonness of Weak Resistance Genes   160
7.10 An Unsolved Geographical Problem   161
7.11 Four Arbitrary Categories of Adaptation in the Pathogen to the
     Host. Abundant Preexisting Virulence. Virulence by Adaptation.
     Restricted Virulence. Forbidden Virulence   162
7.12 The Need to Study Populations in Order to Estimate Fitness. The
     Error of Typifying Populations by Single Isolates   164
7.13 Horizontal Resistance. Absence of Differential Interaction   166
7.14 Anthracnose of Beans. Examples of Physiological Reactions Involved
     in Vertical and Horizontal Resistance   167
7.15 The Commonness of Horizontal Resistance. Horizontal Resistance
     through Normal Metabolic Processes   170
7.16 Vertical versus Horizontal Resistance. New Genes versus old Genes.
     Oligogenic versus Polygenic Resistance. Gene Diversification versus
     Gene Duplication. A Molecular Theory of Horizontal Resistance   172
7.17 The Inequality of Horizontal Resistance Genes   174
7.18 The Quantity of Horizontal Resistance. The Harmfulness of Excess
     of Horizontal Resistance   176
7.19 Immunity as Extreme Resistance   181
7.20 Antigens of Host and Parasite in Vertical Resistance and Immunity   182
7.21 Immunity against Viruses   184
7.22 Theory of Immunity or Vertical Resistance through Phytoalexins or
     Hypersensitivity a Genetic Misfit   185
7.23 An Interlude on Resistance against Secondary Infection. Preformed
     Localized Resistance. A Role for Phytoalexins and Hypersensi-
     tivity   187
7.24 Immunity Compared with Vertical Resistance   190
7.25 Virulence and Aggressiveness   192
7.26 Shared Phenotypic Effects of Three Genotypes   192
7.27 Some Comments about Terminology   193

BIBLIOGRAPHY   197

SUBJECT INDEX   211

Preface

The pathogen, the host plant, the environment, time, and space determine degree of infection. This book discusses these factors and how they interact. The principles of plant infection are presented as an integrated science. The emphasis is on unification and generalization. With few exceptions, the material discussed applies to infection by fungi, bacteria, or viruses, as well as to infection of root, stem, or leaf. New ideas are developed, scattered data collated, and old concepts dropped.

The existing literature on plant pathology is concerned largely with epidemic disease. The special features of endemic disease are neglected by comparison. To help remedy this, a chapter on endemic disease is included. Endemic disease, considered in terms of host–pathogen relationships, is a climax ecological association; and for obligate parasites at least the climax is necessarily unstable except for the special case of tolerance by the host of the pathogen. In botanical ecology there are few fields likely to reap greater rewards than the study of endemic disease, with the pathogen as the focus of study.

Theories of host resistance to disease have been developed considerably in this book. Chapter 7 stresses the dual nature of resistance; and small beginnings of a molecular theory for both types of resistance are suggested. Just as some may think the suggestions farfetched, others are soon likely to assert that the suggestions were obvious all along.

The book deals primarily with quantitative relationships, and recognizes that plant pathology is, or should be, a quantitative science. The need for quantitative studies has grown with the development of fast and capacious computers that allow programs to be written which simulate the increase and spread of disease. The subjects discussed in this book can be easily adapted for use in computer programming.

Quantitative studies also reveal gaps in our knowledge. A clear ex-

ample concerns sites susceptible to infection, surely among the most elementary of topics of plant infection. An analysis of disease/inoculum
curves suggests a two-path concept of susceptibility. Yet we do not
know why even with the same disease susceptible sites are sometimes
numerous but individually not easily infected, whereas others are easily
infected but are relatively few. Differences in the relative importance of
the number of susceptible sites and their potential for infection occur
not only when the pathogen enters through wounds (for then an explanation might seem deceptively easy), but also when it enters through
natural openings such as stomata or through a seemingly intact cuticle.
These differences are of greatest importance in quantitative epidemiology, because all existing mathematical models of epidemics assume,
often incorrectly, that susceptible sites in healthy tissue are inexhausti-
ble. In addition, any basic probe into the nature of susceptibility must at
some stage distinguish between the quality and quantity of the susceptible sites.

If only those pathogens that survive primarily as a result of their
parasitism are considered and pathogens such as *Sclerotium rolfsii* that
live largely as saprophytes and scavengers are ignored, a most remark-
able feature is the specificity of the host–parasite interaction. For any
particular parasite some plants act as hosts, others do not. Evidence
that the explanation of this specificity must be sought at the nucleic acid
level, or, at least, at the level of primary coded polypeptides is mount-
ing. At present this represents the greatest gap in the biochemistry and
physiology of infection processes. Recent research has supplied a new
tool that may help fill this gap: The naked strands of infectious nucleic
acid known as viroids or metaviruses, the smallest known pathogens,
infect without many of the encumbrances that necessarily complicate
infection when the genetic apparatus of parasitism is more complicated.
Theirs is parasitism at its simplest.

I retain the term obligate parasite even for those fungi, such as
Puccinia graminis, that have been grown on artificial culture media.
it has been demonstrated that these fungi grow saprophytically in nature
at some phase, I must regard growth in axenic culture as being an
artifact irrelevant to natural infection. It is to natural infection that I
apply the term obligate parasite.

As we become more engrossed in our own narrow specialities, we
become more likely to mistake an artifact for the real thing. Unquestionably, artifacts can serve a valid purpose; but this does not include
evading a study of natural processes that involve substantial technical

difficulties. To stem the drift toward irrelevant artificialities we need better technologies for collecting experimental data and better analytical probes for handling field observations. Technology is outside the scope of this book, but analysis is the theme of most of its chapters.

My grateful thanks are due to Professor William Merrill for information about the distribution of oak wilt disease and to Dr. J. C. Mooi for information about the occurrence of virulence in *Phytophthora infestans* in the Netherlands.

J. E. VAN DER PLANK

*Principles of Plant
Infection*

Chapter 1

The Relation between the Amount of Inoculum and the Amount of Disease It Produces

1.1 THE QUESTION

What is the quantitative relation between the amount of inoculum and the amount of disease it produces, that is, between inoculum dose and disease response? If 1000 spores fall on a plant and start n lesions, how many lesions would 2000 spores have started? Answers to questions of this sort are the sole topic of Chapters 1 and 2.

Chapter 1 begins by outlining the problems to be discussed. In this outline fungus diseases are used for illustration. The outline is followed by a more detailed inquiry, with fungus, bacterial, and virus diseases dealt with in turn.

Chapters 1 and 2 differ in that they concentrate on opposite ends of the disease/inoculum curve. Chapter 1 deals primarily with disease/inoculum relations when the amount of disease is relatively low, and Chapter 2 with these relations when the amount of disease is relatively high. The problems encountered in the two chapters are essentially different.

In Chapters 1 and 2 things other than the amount of inoculum are assumed to remain equal. That is, these chapters are concerned with the amount of inoculum as the only independent variable. Discussion of the vast array of other factors that affect disease—factors as diverse as the humidity of the air, the presence of pollen, or the susceptibility of the host plants—is left to later chapters.

1.2 THE SIMPLEST RELATION: DISEASE IS PROPORTIONAL TO INOCULUM

Figure 1.1 shows disease directly proportional to inoculum. It reproduces some data of Schein (1964) on bean rust. Uredospores of *Uro-*

1

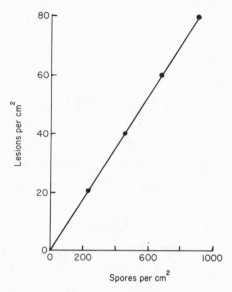

Fig. 1.1   The relation between the number of lesions per cm² of leaf of *Phaseolus vulgaris* and the number of spores of *Uromyces phaseoli* per cm². (Data of Schein, 1964.)

myces phaseoli were suspended in dilute agar and sprayed on half-leaves of *Phaseolus vulgaris*. On the *x*-axis is shown the number of spores deposited on each cm² of leaf surface, and on the *y*-axis the number of lesions that developed per cm² in due course.

Consider two features of Fig. 1.1. First the line fitted to the points is straight (or, to be precise, there is no evidence of significant departures from straightness). Second, the line when extended passes through the origin of the graph: the point (0,0) where the *x*- and *y*-axes meet. These two features taken together mean that the number of lesions is proportional to the number of spores, the ratio of lesions to spores being about 1:11 at all points.

Figure 1.1 represents infection in which the spores neither interact nor compete noticeably with one another for susceptible sites on the leaf.

1.3   THE COMMONEST RELATION: THE RATIO OF DISEASE TO INOCULUM DECREASES AS THE AMOUNT OF INOCULUM INCREASES

Figure 1.2 shows a common relation. The disease/inoculum curve is no longer straight, as in Fig. 1.1, but curves to the right. With more

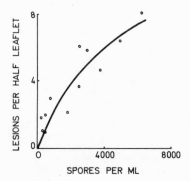

Fig. 1.2   The relation between the number of lesions per half-leaf of *Vicia faba* and the number of spores of *Botrytis fabae* per milliliter of inoculum. (Data of Last and Hamley, 1956.)

spores in the inoculum, the ratio of lesions to spores decreases. The data are those of Last and Hamley (1956) for the number of lesions developed on leaves of *Vicia faba* when conidial suspensions of *Botrytis fabae* were rubbed on them. (For further comment on these data see Section 2.9.)

Figure 1.2 represents infection without interaction between spores but with competition for susceptible sites on the leaf. There is more about this in Chapter 2.

1.4   THE EFFECT OF ANTAGONISTIC INTERACTION BETWEEN SPORES: AS INOCULUM INCREASES DISEASE INCREASES TO A MAXIMUM AND THEN DECREASES

Figure 1.3 is based on data of Davison and Vaughan (1964). It shows the effect of antagonistic interaction between uredospores of *Uromyces phaseoli*. Bean leaves were inoculated with suspensions of spores, and after appropriate incubation the number of lesions was counted. Figure 1.3 relates the number of lesions to the number of spores per cm^2 of leaf surface. With increasing inoculum the number of lesions increases to a maximum of about 27 lesions per cm^2. Thereafter, with still further increases of inoculum, the trend is reversed and the number of lesions decreases, fewer lesions being formed by 3400 than by 1100 spores/cm^2.

Inhibition of germination probably accounts for fewer lesions being produced when inoculum is abundant.

Self-inhibition of germination seems to be common in the uredospores of the rust fungi generally (Yarwood, 1954; Wilson, 1958; Bell

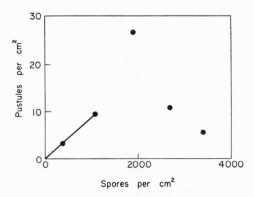

Fig. 1.3 The relation between the number of pustules per cm^2 of leaf of *Phaseolus vulgaris* and the number of spores of *Uromyces phaseoli* per cm^2. (Data of Davison and Vaughan, 1964, from Van der Plank, 1967a, p. 68.)

and Daly, 1962; Woodbury and Stahmann, 1970; Marte, 1971; and Macko *et al.,* 1972). The topic has also been reviewed by Allen (1965) and Wood (1967).

Inhibitors of germination occur in spores of other fungi as well. Thus Domsch (1953) obtained a disease/inoculum curve for *Erysiphe graminis* on barley similar to that of Davison and Vaughan for *Uromyces phaseoli* on bean. It seems likely that the inhibition of germination after spores have been dispersed, as distinct from inhibition before they are dispersed, is an artifact. This is discussed in Section 2.4.

The difference between Figs. 1.2 and 1.3 is the difference between an asymptote and a maximum. In Fig. 1.2 disease increases with increasing inoculum without ever decreasing. In Fig. 1.3 disease increases with increasing inoculum until a maximum has been reached, and thereafter declines.

1.5 THE EFFECT OF SYNERGISTIC INTERACTION BETWEEN SPORES: AS INOCULUM INCREASES IT BECOMES INCREASINGLY EFFECTIVE

Figure 1.4 shows the effect of synergistic interaction between uredospores of *Puccinia graminis*. Petersen (1959) allowed uredospores to fall on wheat plants. After an appropriate interval in an infection chamber and greenhouse he stained the leaves and counted the infection points. The counts are recorded in Fig. 1.4. The first nine points, up to 2810 spores and 330 infection points/cm^2, are more or less in a

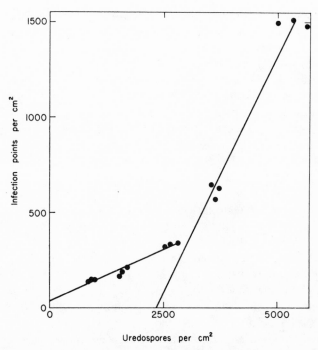

Fig. 1.4 The relation between the number of infection points per cm² of wheat leaf and the number of uredospores of *Puccinia graminis tritici* per cm². (Data of Petersen, 1959, from Van der Plank, 1967a, p. 66.)

straight line and there is no evidence of synergistic interaction at this stage. The number of infection points stayed roughly constant at about 12 per 100 spores. But beyond this stage, as the number of spores was increased, there was a more than proportional increase in the number of infection points until with 5400 spores/cm² there were 27 infection points per 100 spores. At this high concentration, the spores helped one another to infect; they acted synergistically. There is other evidence for this. Petersen (1959) found that high concentrations of spores stimulated germination. With only 850 spores/cm², 28% of the spores germinated and produced appressoria; with 5300 spores/cm², 52% germinated and formed appressoria.

The production by spores of germination stimulants, particularly the higher straight-chain aldehydes and alcohols, has been extensively studied. The literature has been reviewed by Allen (1965) and Wood (1967).

In relation to synergism, the work of Rapilly (1968), Rapilly and Fournet (1968), Rapilly *et al.* (1970), and Stanbridge and Gay (1969) introduces the concept of units of dispersal into epidemiology. Uredospores of *Puccinia striiformis* are dispersed in clumps, the uredospores being held together by a mucilaginous coating. Within these clumps spores germinate better than when they occur singly; but the number of uredosori formed is independent of whether infection starts from a single germ tube or several in the same clump. The synergistic interaction between spores in the clump—the unit of dispersal—allows spores to infect better. The matter is taken up again in Section 1.20.

1.6   AN UNDEMONSTRATED RELATION: A NUMERICAL THRESHOLD OF INFECTION

In Figs. 1.1–1.4 curves were considered that, within the limits of experimental accuracy, pass through the origin: the point (0,0). We must now consider curves that do not.

Gäumann (1964), Sadasivan and Subramanian (1960), Nutman and Roberts (1963), Carter (1972), and others have claimed that a minimum number of spores greater than one is necessary, even in favorable conditions, to establish certain diseases. This is the numerical threshold of infection. Thus, according to Gäumann, 200 resting spores of *Synchytrium endobioticum* per gram of soil are the minimum necessary to cause wart disease of potatoes. Anything less is ineffective. An infection spot caused by a single myxamoeba does not lead to infection, because the morphogenic stimulus is too weak. Similarly, with bunt of wheat, 100 spores of *Tilletia* per grain of wheat are the minimum needed to infect even the very susceptible variety Jenkins' Club, and as many as 500 to 5000 to infect the less susceptible variety Marquis.

Both this and the previous section are concerned with synergism and its effect on disease/inoculum curves. Single spores of *Puccinia graminis,* our topic previously, can infect—that has been demonstrated over and over again in single-spore inoculations—but spores in mass help one another to infect even better. Synergism here is facultative. But the theory of a numerical threshold, our topic in this section, denies for many fungi the possibility of a single spore infecting. A threshold number is needed. Synergism, on this theory, is obligate. This is not the place to compare and contrast facultative and obligate synergism. All that is relevant for the present is to note that the two forms of synergism give very different disease/inoculum curves. Facultative synergism, shown in Fig. 1.4, markedly affects the disease/inoculum curve at high levels of disease, but is without noticeable effect at low levels, i.e., near the origin of the graph. Obligate synergism, shown in Fig. 1.5, markedly affects the disease/inoculum curve at low levels of disease.

Figure 1.5 uses Gäumann's statements about *Synchytrium endobioticum* for illustration. As in Figs. 1.1–1.4, the scales are arithmetic. On the x-axis is the number of resting spores per gram of soil, on the y-axis the number of lesions (warts) per potato plant. The x-axis is also the line $y = 0$, i.e., the line of no disease. According to Gäumann there is no disease with less than 200 spores per gram of soil, so the disease/ inoculum line starts from the x-axis far to the right of the origin, at the point corresponding to $x = 200$ spores per gram. Thereafter, the line curves upward, this curvature corresponding to the synergistic interaction between myxamoebae or infection spots which Gäumann assumed.

Nothing like the disease/inoculum curve in Fig. 1.5 has ever been recorded experimentally. Neither the experimental data for *S. endobioticum* nor the data for all the other organisms cited by Gäumann and his followers support the theory of a numerical threshold of infection. These data will be considered in much detail in Section 1.10. For the present we are concerned only with outlining the contents of this chapter. But before proceeding to the next type of disease/inoculum curve, two other points need to be made about the theory of a numerical threshold of infection.

There are two ways of thinking of the threshold. There can either be an environmental threshold or a threshold at the susceptible site. Gäu-

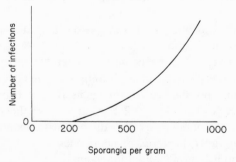

Fig. 1.5 The interpretation of Gäumann's (1946) theory of a numerical threshold of infection, which requires a minimum of 200 sporangia of *Synchytrium endobioticum* per gram of soil before infection can occur.

mann specified a threshold in the soil environment when he specified that a minimum of 200 resting spores of *S. endobioticum* per gram of soil were needed before potatoes would become infected with wart disease. Figure 1.5 interprets this environmental threshold. The other sort of threshold, a threshold at the susceptible site, is illustrated in Fig. 1.17. This shows a disease/inoculum curve calculated on the assumption that at least 10 randomly scattered particles must interact synergistically at a susceptible site before infection can begin there. The curve bends characteristically upward after starting at the origin and hugging the x-axis for some distance. This curve too does not fit the known facts. (Like Gäumann and his followers, we are discussing fungus disease. The story of some viruses is different, and will be told later.)

Gäumann's reference to wart disease of potatoes suggests that he saw the disease process in two stages: first, the formation of infection spots by the myxamoebae, and, second, the synergistic interaction between adjacent infection spots to form lesions. Wood (1967) supports him, and believes that the combined action of adjacent infection spots to form a lesion is more likely than the combined action of adjacent spores to form an infection spot. The distinction between spores and infection spots does not blunt our criticism. Synergistic interaction, irrespective of whether it occurs early between spores or later between infection spots, would produce the same sort of disease/inoculum curve; and if this sort of curve does not occur, synergistic interaction does not occur at any stage.

1.7   AN IMPOSSIBLE RELATION: INFECTION WITHOUT INOCULUM.
EVIDENCE AGAINST RECURRENT HETEROGENESIS OF VIRUSES

The relation has no interest nowadays for fungi and bacteria apart from its geometric implications.

Three hundred years ago Francesco Redi refuted the idea that maggots arise spontaneously when meat decomposes. He covered meat with a layer of muslin, and showed that maggots do not occur unless flies could lay their eggs on it. In 1862 Pasteur carried the matter further and proved that microorganisms also do not arise spontaneously. For disease caused by fungi, bacteria, or nematodes the issue has long been settled: there is no infection without inoculum.

The geometric implication is that the disease/inoculum curve shown in Fig. 1.6 is impossible. A disease/inoculum curve cannot cut the y-

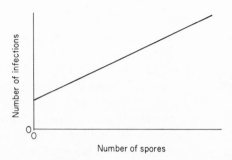

Fig. 1.6   A disease/inoculum curve cutting the y-axis above the origin is impossible, because it implies the spontaneous generation of disease, i.e., disease without inoculum.

axis above the origin. The y-axis is also the line $x = 0$, i.e., the line for zero inoculum; and a curve cutting the y-axis above the origin is a curve for disease without inoculum.

If one finds a disease/inoculum curve cutting the y-axis above the origin, there can be three explanations. First, the curve may be due to sampling errors, i.e., the deviation from the origin is not statistically significant. Second, the result may be due to faulty curve fitting, especially to trying to fit a straight line to points that are not in a straight line. For example, if a straight line were fitted to the points in Fig. 1.2 it would cut the y-axis above the origin. Third, the result may be due to inoculum undetected by the experimenter, i.e., to undetected contamination.

Viruses are different; and Semancik and Weathers (1972) have revived the theory of spontaneous generation. The viruses involved are those of tomato bunchy top, potato spindle tuber, and citrus exocortis. These viruses are similar and possibly identical. Their special interest to virologists is their very low molecular weight, probably much too low for them to be able to code for their own replication (see Section 1.17). Semancik and Weathers suggest that they may be "pathogenes" or constitutive host ribonucleic acid which depends for its multiplication on normal host synthetic pathways but at the same time becomes autonomous and simulates symptoms caused by pathogens. They are, in fact, suggesting the heterogenesis of viruses, because by their easy transmissibility from plant to plant these autonomous agents of disease are clearly viruses by all current definitions. Various names for these agents such as viroid or metavirus have been suggested; but they make distinctions without a difference to this paragraph. Tomato bunchy top and potato spindle tuber viruses are easily transmitted mechanically, and potato spindle tuber and citrus exocortis viruses by vegetative propagation.

The evidence of geographical distribution is against the pathogene theory, if it implies recurrent heterogenesis. Tomato bunchy top has been recorded only from southern Africa. Its symptoms are conspicuous. If the virus were generated spontaneously and recurrently, the disease would surely have been recorded peppering the millions of acres of tomatoes in other continents. So too potato spindle tuber has been recorded only from North America, though this evidence is less strong because symptoms are often vague and might well have been missed elsewhere.

This evidence does not dispute the possibility that these viruses might have started as runaway fragments of host nucleic acid which have since been replicated by host synthetic pathways. About this we have no evidence and make no comment. But the evidence does dispute the possibility of heterogenesis being a recurrent process, which is to say the evidence shows that for viruses as well as fungi and bacteria Fig. 1.6 is impossible.

Diener (1973) points out that viroids, that is, infectious ribonucleic acids, may be distinct from viruses. We need not take sides here about this, because we can couple viruses and viroids equally in denying that evidence exists for recurrent heterogenesis.

1.8   A SUMMARY. THE GENERAL PROBLEM: WHAT CHAPTERS 1 AND 2 ARE ABOUT

Figure 1.7 summarizes the six relations discussed in previous paragraphs. As before, the scale is arithmetic, and both inoculum on the x-axis and disease on the y-axis start from zero. Curve A is for disease proportional to inoculum, as in Fig. 1.1. Curve B is for a decreasing ratio of disease to inoculum, as in Fig. 1.2. Curve C is for antagonistic interaction between spores, as in Fig. 1.3. Curve D is for facultative synergism between spores, as in Fig. 1.4. Curve E is for the unproven theory of a numerical threshold of infection, as in Fig. 1.5. Curve F represents the impossible relation of disease without inoculum, as in Fig. 1.6.

Our problem is this. Of all these curves, only Curve A is assumed in

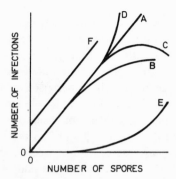

Fig. 1.7 Six disease/inoculum curves. Of these, four—*A, B, C,* and *D*—are known, one—*E*—is unknown, and one—*F*—is impossible.

general models of epidemic disease. The widely accepted theory that disease increases exponentially (or geometrically or logarithmically or at compound interest, to give synonyms) can be valid only if Curve *A* is valid, i.e., only if the number of lesions is proportional to the number of spores. Models of epidemics all assume the relation in Curve *A*. This is true not only for plant pathology but for medical epidemiology as well. Bailey's (1957) "Mathematical Theory of Epidemics" assumes throughout that (with due allowance for the working of the laws of chance) disease is directly proportional to inoculum. To what extent are these theories and models soundly based? With what justification can we ignore Curves *B, C,* and *D* which after all represent experimental fact?

Changing from Curve *A* to Curves *B, C,* or *D* would affect epidemics mainly in their later stages after disease had mounted. If Curve *B* held instead of *A* epidemics would tend to slow down as they proceeded. Actually it is not uncommon for epidemics to slow down more in their later stages than one would expect from models that assume Curve *A*. If Curve *C* held, the slowing down would be greater; epidemics might even grind to a halt. If Curve *D* held, epidemics would become increasingly explosive as they proceeded, an effect not observed except when there is some obvious reason such as increasing susceptibility of the host plants as they grow older or an environment increasingly favorable to disease as the season advances. If Curve *E* (for a numerical threshold of infection) held, epidemics would be affected right from the start; epidemics caused by obligate parasites such as *Synchytrium endobioti-*

cum could not easily begin from small beginnings, and the pathogen would be burdened by its obligate synergism. See Section 1.18.

In Chapters 1 and 2 we discuss disease/inoculum curves in more detail and more critically than others have attempted to do before. The previous paragraphs put the detail in perspective. Only by distinguishing between the various disease/inoculum relations and understanding them can we understand the process of disease in field, forest, or orchard. What is probed in Chapters 1 and 2 is the keystone to quantitative studies of how disease increases—or decreases or stays static—in nature; and no apology is needed for the amount of detail in our probe.

1.9   MULTIPLE INFECTION OF PLANTS: THE PROBLEM OF ASSESSING DATA ABOUT THE PERCENTAGE OF DISEASED PLANTS

Figures 1.1–1.4 are based on actual counts of the number of lesions or infection points in relation to the amount of inoculum. More commonly, plant pathologists give information about the relation between inoculum and disease in terms of the percentage of diseased plants. If disease occurs as separate and visible lesions, the statement that a plant was diseased does not tell us whether one lesion was found on it or a thousand. If disease is systemic, the statement that a plant was diseased does not tell us how often inoculum was separately introduced into the plant. An oat seedling, for example, can be infected more than once, independently, by smut. This was neatly proved by Person and Cherewick (1964) for *Ustilago kolleri* and *U. avenae*. With genetic markers they showed that as many as four separate infections can participate in developing disease in a single oat plant. So, too, Meyer and Maraite (1971) produced a composite systemic infection in melon plants by inoculating them simultaneously with two races of *Fusarium oxysporum* f. sp. *melonis*. With bacteria, multiple infection can be inferred from the experiments of Ercolani (1967). He isolated two variants of *Corynebacterium michiganense,* one being resistant to terramycin and the other susceptible. This was the only demonstrable difference between the variants, which were equally virulent in tomato plants and had the same growth rate *in vivo* and *in vitro*. He used the difference in response to terramycin as a "marker" character, and showed that when tomato plants were inoculated with concentrated suspensions of the two variants, in excess of 1 ED_{50}, the ratio between the two was much the same in the infected tomato plants as in the inoculum used to infect them. (These results are not to be confused with those for weak suspensions discussed in Section 1.13.) With viruses also, multiple infection can reasonably be assumed to occur in systemic disease. The process is obvious with viruses that start by producing local lesions at the infection points and then become systemic.

To study quantitative relations between inoculum and infection it is necessary to convert the percentage or proportion of plants diseased into the estimated number of lesions or infection points per plant. Estimates can be made by assuming that the lesions or infection points are randomly distributed over the plants. Estimates are only as good as the assumption of randomness; and it will be clearly stated throughout this book whether a graph or table is based on observed numbers of lesions or on estimated numbers.

The estimated number is $-\log_e(1-y)$, where y is the proportion of diseased plants (and $100y$ the percentage of diseased plants). For example, if $y = 0.2$, then $-\log_e(1-y) = 0.223$, which is the estimated average number of infections per plant. That is, an average of 20% of the plants will be infected when 223 infections are randomly scattered among every 1000 plants. Of these 223 infections, 23 will be second or third or fourth . . . infections of plants already infected once.

The principle is this. When infections are distributed randomly among homogeneous plants, the probability of any one plant remaining healthy is e^{-m}, where m is the mean number of infections per plant. But when the number of plants is large the probability of a plant being healthy is numerically the proportion of healthy plants:

$$e^{-m} = 1-y \qquad m = -\log_e(1-y) \qquad (1.1)$$

A table of $-\log_e(1-y)$, written as $\log_e[1/(1-y)]$, is given by Van der Plank (1963, Appendix, Table 3).

Two comments are needed. First the estimated mean number of infections m is used in graphs in the same way as the observed number was used in the previous graphs. The scale is purely arithmetic, even though logarithms are used as a vehicle to calculate m. Second, the method is accurate at low percentages of disease. Indeed, it scarcely matters at this stage whether one plots the observed proportion (or percentage) of disease or m. See Table 1.1. In the next few sections the argument is concerned with the behavior of curves near the origin of the graph, i.e., with low proportions of disease; and the transformation is adequate.

TABLE 1.1

Values of $m = -\log_e(1 - y)$ [a]

y	m	y	m	y	m
0.01	0.010	0.17	0.186	0.46	0.616
0.02	0.020	0.18	0.198	0.48	0.654
0.03	0.030	0.19	0.211	0.50	0.693
0.04	0.041	0.20	0.223	0.55	0.799
0.05	0.051	0.22	0.248	0.60	0.916
0.06	0.062	0.24	0.274	0.65	1.050
0.07	0.073	0.26	0.301	0.70	1.204
0.08	0.083	0.28	0.329	0.75	1.386
0.09	0.094	0.30	0.357	0.80	1.609
0.10	0.105	0.32	0.386	0.85	1.897
0.11	0.117	0.34	0.416	0.90	2.303
0.12	0.128	0.36	0.446	0.92	2.526
0.13	0.139	0.38	0.478	0.94	2.813
0.14	0.151	0.40	0.511	0.96	3.219
0.15	0.163	0.42	0.545	0.98	3.912
0.16	0.174	0.44	0.580	0.99	4.605

[a] Here y is the proportion of plants infected and m the estimated mean number of infections per plant. This table illustrates how close m and y are at the beginning and how they diverge later. For a fuller table see Van der Plank (1963), pp. 320–322.

1.10 THE THEORY OF A NUMERICAL THRESHOLD OF INFECTION. EVIDENCE AGAINST IT. CONFUSION WITH A DILUTION END POINT

The theory of a numerical threshold of infection was introduced in Section 1.6, and rejected on the grounds that the available experimental evidence contradicts it. This evidence was not then examined in detail. The time has come to do this now.

The theory requires a disease/inoculum curve of a sort which has never been found for fungus diseases, and cannot be derived from the data that Gäumann (1946) and his followers themselves cite as evidence. It seems fair to concentrate on these data, all for fungus disease, and show that they prove the opposite of what they were said by Gäumann to prove. This situation has arisen from a confusion between two separate concepts: a numerical threshold of infection and a dilution end point.

Figure 1.5 interpreted Gäumann's (1946) statements about wart disease of potatoes caused by *Synchytrium endobioticum*. The disease/inoculum curve starts from the x-axis at the point corresponding to 200 resting spores per gram of soil, representing Gäumann's statement that the disease cannot occur in soil with less than 200 resting spores per gram. And the curve bends upwards, representing Gäumann's statement that an infection spot caused by a single myxamoeba cannot lead to infection, because the morphogenic stimulus is too weak, i.e., more than one myxamoeba or infection spot is needed. (The relation between the interaction of spores or infection spots at a susceptible site and the curvature of the disease/inoculum curve is discussed in Section 1.12.)

Figure 1.8 reproduces Glynne's (1925) data on *S. endobioticum* on which Gäumann based his statement. Figure 1.5 and 1.8 are totally unlike, and curve in opposite directions. Although Gäumann used it, Fig. 1.8 is not altogether a fair test of his theory. It would be fairer to transform the percentage of diseased plants into an estimated number of lesions per plant, using the method given in Section 1.9. Figure 1.9 does this; it plots the estimated mean number of infections per plant against the number of resting spores per gram of soil. Within the limits of experimental error, the line fitted to the points is straight and passes through the origin. (The line was fitted by standard statistical procedure to avoid any subjective bias.) We are, in fact, back at a representation of simple proportionality between inoculum and disease, as previously illustrated by the bean rust fungus *Uromyces phaseoli* in Fig. 1.1 or Curve *A* in Fig. 1.7.

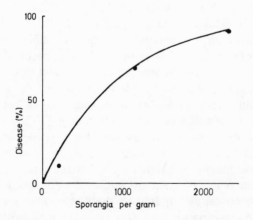

Fig. 1.8 The relation between the percentage of diseased potato tubers and the number of sporangia of *Synchytrium endobioticum* per gram of soil. (Data of Glynne, 1925.)

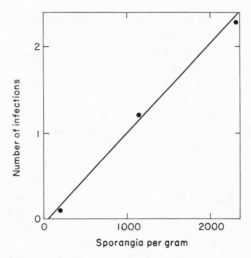

Fig. 1.9 The relation between the estimated number of infections per tuber and the number of sporangia of *Synchytrium endobioticum* per gram of soil. The number of infections is estimated as *m* in Eq. (1.1) Figures 1.8 and 1.9 should be compared to show the effect of allowing for multiple infections by Eq. (1.1). (Same data as in Fig. 1.8, modified from Van der Plank, 1967a, p. 69.)

Other evidence cited by Gäumann in support of his theory is repro-duced in Figs. 1.10–1.12; that cited by Nutman and Roberts (1963) in Fig. 1.13; and that cited by Carter (1972) in Fig. 1.14.

Figure 1.10 reproduces the data of Heald (1921). It shows the amount of bunt in wheat plants of the very susceptible variety Jenkins' Club grown from grain artificially infected with spores of *Tilletia* spp. Only data for up to 40,000 spores per grain are included; from Heald's survey it is known that higher loads are abnormal on grain used for seed, and Heald's data for higher loads would not affect our conclu-sions. Figure 1.11 reproduces data of Haymaker (1928) on wilt of to-matoes caused by *Fusarium oxysporum* f.sp. *lycopersici*. In neither Fig. 1.10 nor 1.11 is there any suggestion of the upward curvature of the disease/inoculum line required by Gäumann's theory, nor do the lines fitted to the points by standard statistical methods cut the *x*-axis to the right of the origin, as required by the theory. (Within the limits of much experimental variation, the data are compatible with simple pro-portionality between inoculum and disease, as in Fig. 1.1 or Curve *A* of Fig. 1.7.) Figure 1.12 reproduces the data of Dickson (1923) on seed-ling blight of maize caused by *Gibberella saubinetii*. (The data are for

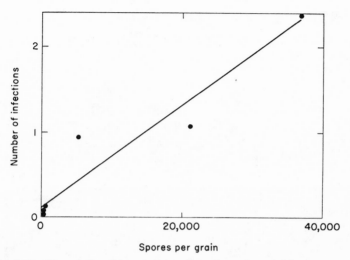

Fig. 1.10   The relation between the estimated number of infections per wheat plant and the number of spores of *Tilletia* spp. per grain of seed wheat. The number is esimated as *m* in Eq. (1.1). (Data of Heald, 1921, modified from Van der Plank, 1967a, p. 70.)

combined preemergence blight and severe wilt.) The curvature is the opposite of what the theory of a numerical threshold requires. (The curve is the same as the curve in Fig. 1.2 or Curve *B* of Fig. 1.7. Increasing the number of spores decreases the proportion that infect.) In Figs. 1.10–1.12 the data are transformed into the estimated mean num-

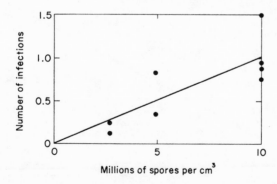

Fig. 1.11   The relation between the estimated number of infections per tomato plant and the number of spores of *Fusarium oxysporum* f. *lycopersici* per cm³ of inoculum. The number is estimated as *m* in Eq. (1.1). (Data of Haymaker, 1928, modified from Van der Plank, 1967a, p. 74.)

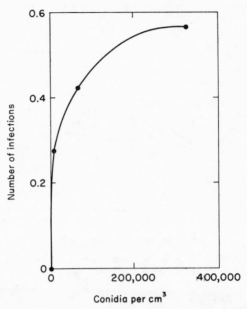

Fig. 1.12 The relation between the estimated number of severe infections per maize plant and the number of conidia of *Gibberella saubinetii* per cm³ of inoculum. The number is estimated as *m* in Eq. (1.1). (Data of Dickson, 1923, modified from Van der Plank, 1967a, p. 74.)

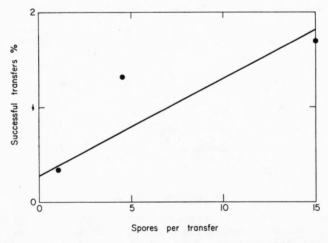

Fig. 1.13 The relation between the percentage of successful transfers to coffee leaves and the number of spores of *Hemileia vastatrix* per transfer. (Data of Nutman and Roberts, 1963, from Van der Plank, 1967a, p. 75.)

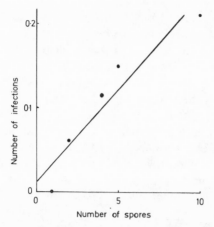

Fig. 1.14   The relation between the estimated number of infections on peach leaves and the number of spores of *Tranzschelia discolor* per drop of inoculum. The number is estimated as m in Eq. (1.1). (Data of Carter, 1972.)

ber of infections per plant. The untransformed data for percentages of disease, used by Gäumann, accord even worse with his theory. In Fig. 1.13 other data are presented without transformation; the percentages are so low that transformation would have no visible effect on the graph. Figure 1.13 reproduces the data of Nutman and Roberts (1963), who inoculated coffee leaves with spores of the rust fungus *Hemileia vastatrix*. They used varying numbers of spores in drops of water, and counted the percentage of successful transfers. The data are not very precise; 10 to 20 spores per drop is interpreted as 15 spores per drop. The straight line fitted to the points by standard statistical procedure cuts the y-axis above the origin, i.e., if projected the line would cut the x-axis to the left of the origin. (This departure from the origin is statistically insignificant.) There is no hint here of evidence for the theory of a numerical threshold. Figure 1.14 reproduces the data of Carter (1972), who inoculated peach leaves with spores of the rust fungus *Tranzschelia discolor* in drops of water and counted the proportion of inoculated sites on which sori developed. Again, the line fitted to the points cuts the y-axis above the origin, and there is no evidence for the threshold theory.

In all these examples the authors have confused a numerical threshold of infection with a dilution end point. A dilution end point, used extensively in virology, is the greatest dilution of inoculum compatible with being reasonably sure of getting infection. Thus, virologists are agreed that in ordinary conditions a tobacco plant must be inoculated with about a million particles of tobacco mosaic virus for it to be probable that the plant will become infected. That is, the dilution end point of a solution of tobacco mosaic virus is such that the solution applied to the plant contains about a million particles of the virus. But virologists are equally agreed that the actual infection is by a single particle: by only one out of the million applied to the plant, the rest taking no part. The disease/inoculum curve for tobacco mosaic is what virologists call a "one-hit" curve. See Section 1.15. So too one can fully agree with Gäumann that to be reasonably sure of infecting a potato plant with wart disease one should grow the plant in soil containing at least 200 resting spores per gram. That is, 200 resting spores per gram is the dilution end point of *S. endobioticum* in soil. But the actual infection is carried out by a single myxamoeba without help or hindrance from all the other thousands in the soil. The disease/inoculum curve in Fig. 1.9 is a "one-hit" curve. The probability that any one particular myxamoeba will start infection is small. If and when it starts infection, it does it alone. The last two sentences deal with entirely separate concepts.

Even some of the experts in the field are confused. As an example of the unwarranted association of synergism with the ED_{50}, consider the following quotation. Garrett (1970) analyzed Heald's (1921) data about *Tilletia* spores on wheat grains and wrote that "the ED_{50} must have been at least 9×10^4 spores, so it is not surprising that Heald should have concluded that synergism was essential for infection." A high ED_{50} means no more than a low probability that a given single spore will infect; and the quotation illustrates a common reluctance to accept the fact that nature is often exceedingly prodigal in reproductive processes. This prodigality is not a special feature of the infection process in plant disease, but extends throughout the vegetable and animal kingdoms. A few examples are given in Section 1.21.

The evidence of all known disease/inoculum curves is consistent: There is no known obligate synergism between spores or sporelings in infection by fungi.

1.11 THE EXPERIMENTAL LAW OF THE ORIGIN. THE NATURE OF EXPERIMENTAL LAW

The evidence we have considered shows that the disease/inoculum curve will not cut the *x*- or *y*-axis anywhere except at the origin. It will not cut the x-axis to the right of the origin, because that implies a numerical threshold of infection, which is not known to exist. It will not cut the y-axis above the origin, because that implies the spontaneous generation of pathogens, disbelieved since Pasteur's day. We refer here to axes on an arithmetic scale, and the origin as the point $(0,0)$.

We summarize this as an experimental law: When disease is plotted against inoculum, both on arithmetic scales, the curve starts at the origin.

An experimental law is worth formulating if it gives an accurate enough summary of experimental fact and is useful. Universal accuracy, though desirable, is neither necessary nor implied. Consider the experimental law probably best known to all students of science. Boyle's law states that at constant temperature the volume of a gas varies inversely as the pressure. Boyle formulated this law in 1661 as a summary of his experiments with pressures ranging from 3 to 300 cm of mercury. Later work with more-refined apparatus and with other gases and temperatures has shown that the law is not highly accurate even over this range of pressure, and that the law cannot be extended to high pressures above, say, 50 atm. Nevertheless the law has been useful. It enters into the gas laws used in chemistry and physics. It helped found the kinetic theory of gases. In 1885 van't Hoff transferred it from gases to solutions, and showed both experimentally and theoretically that it applied to osmotic pressures (again with the proviso that it applied only to relatively low pressures, i.e., only to dilute solutions). About the usefulness of Boyle's law there can be no question.

To return to the law of the origin, it is useful. It provides an experimental foundation for models of infection used in epidemiology. It also provides a clear and unambiguous hitching point in disease/inoculum graphs. Here it is incompatible with the cult of relating the amount of disease to the logarithm of the amount of inoculum. These logarithmic or semilogarithmic relations smother essential information. They enter the literature on arguments that are questioned in the next chapter. Here we note their unsuitability for representing the simplest and clearest fact in the whole field of disease/inoculum relations: the law of the origin.

The law of the origin is a useful summary of experience. But, like any other experimental law, it does not predict what may be found in the future, with other pathogens or in other circumstances. Like every experimental law, it is always on probation.

1.12   CURVES DEPICTING SYNERGISM BETWEEN RANDOMLY
DISPERSED SPORES OR INFECTION SITES. RELATIONS
NEAR THE ORIGIN

As was pointed out in Section 1.6, there are two ways of looking at
a numerical threshold of infection. There could be an environmental
threshold, as implied in Gäumann's statement that 200 resting spores of
Synchytrium endobioticum per gram of soil was the minimum concentration of inoculum that would be needed for potatoes to become infected with wart disease. The soil here is the environment. But equally
one can look upon a numerical threshold as the minimum number of
spores that would have to act together at a susceptible site if they were
to cause disease. The experimental evidence for fungus and bacterial
disease is equally against a numerical threshold, whatever way one
looks at it. But we have not yet examined the graphical consequences
of spores acting together at a susceptible site. This is what this section
sets out to do, using the Poisson equations.

Curve *A* of Fig. 1.15 represents single spores or other particles infecting at the susceptible site, without synergism. It is a "one-hit" curve.
It is that part of Curve *B* of Fig. 1.7 that lies very near the origin. For
all practical purposes the line is straight. We shall consider it to be so,
though there is in fact an almost imperceptible curvature to the right
which we shall ignore and which is in the opposite direction to the curvature which we shall now proceed to discuss.

Curve *B* of Fig. 1.15 is for the combined action of two spores or
other particles at the susceptible site. It is for obligate synergism between at least two spores; it is a two-hit curve. The curve starts from
the origin, and curves upward. Figures 1.16 and 1.17 are for synergis-

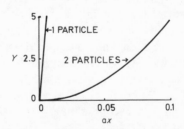

Fig. 1.15 This illustrates two Poisson equations. In the one relation, one particle at a site can on its own start infection; in the other, at least two particles
are needed at a site. Y is the expected number of infections, x the number of
particles, $N = 1000$, and $a = 0.001$.

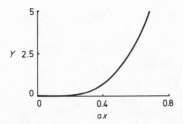

Fig. 1.16 This illustrates the Poisson equation when at least four particles are needed at a site to start infection. Y is the expected number of infections, x the number of particles, $N = 1000$, and $a = 0.001$.

tic interaction between at least four and ten particles, respectively, at the susceptible site. They are four-hit and ten-hit curves, respectively. They also start from the origin, but the upward curvature is more marked, and the ten-hit curve clings quite closely to the x-axis for a while.

The two sorts of numerical threshold of infection differ in relation to the origin. The environmental threshold, illustrated by Curve E of Fig. 1.7, cuts the x-axis to the right of the origin. The threshold consistent with synergistic interaction between spores at the infection site is represented graphically by curves that start from the origin itself.

The essential feature common to both sorts of numerical threshold is the upward curvature of the lines as they leave the x-axis. This upward curvature, we must stress, refers to curves near the x-axis, i.e., to curves depicting low percentages of disease. (As disease increases and approaches 100%, the curvature is reversed, but that is a feature common to most curves, irrespective of whether there is synergism or not, and is irrelevant to our present topic.) At low percentages of disease,

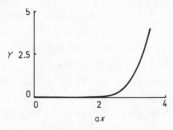

Fig. 1.17 This illustrates the Poisson equation when at least ten particles are needed at a site to start infection. Y is the expected number of infections, x the number of particles, $N = 1000$, and $a = 0.001$.

which we can interpret as less than, say, 5% disease, the obligate synergism demanded by the concept of a numerical threshold of infection will invariably make for an upward curvature, whereas without synergism the line will be almost straight with a practically imperceptible curvature downward. An upward curvature near the origin or x-axis has not been demonstrated for fungus disease; and for this reason we reject the concept of a numerical threshold of infection by fungi, in whatever form the threshold takes.

The Poisson equations used in Figs. 1.15 and 1.17 are as follows. If one spore—or bacterial cell or virus particle—is enough to cause infection (i.e., Curve A of Fig. 1.15) the equation is

$$Y = N\,(1 - e^{-ax}) \tag{1.2}$$

where Y is the number of infections, x the number of spores, and a and N are parameters. This is the one-hit equation. If at least two spores are needed, the equation—the two-hit equation—is

$$Y = N\,[1 - e^{-ax}(1 + ax)]$$

If at least four spores are needed, the four-hit equation is

$$Y = N\,[1 - e^{-ax}(1 + ax + a^2x^2/2! + a^3x^3/3!)]$$

And so on. In the figures we use $a = 0.001$ and $N = 1000$. N is the total number of susceptible sites; and because we are concerned only with low percentages of disease, i.e., with curves near the x-axis, Y/N in Figs. 1.15–1.17 is kept small. The value, $N = 1000$, that is used in the figures is unimportant for our story; the value could be changed without changing the disease/inoculum curves simply by changing the numbers on the y-axis. So, too, the value, $a = 0.001$, is unimportant for our story; the value could be changed without changing the disease/inoculum curves simply by changing the numbers on the x-axis.

The Poisson equations are for randomly distributed spores (or bacterial cells or virus particles). The experimental data in all the figures so far considered are for spores as randomly distributed as the experimenter could make them, so the equations are appropriate. In any case, departure from randomness would not in itself make the curvature upward, which is what we are concerned with.

We have been discussing fungus disease. The same conclusion, that obligate synergism is absent, holds also for bacterial disease, as we shall see in the next section. But it does not hold for all virus diseases, as we shall see later.

1.13 INFECTION BY BACTERIA

Trevan (1927) suggested for bacteria the counterpart of Gäumann's theory for fungi of a numerical threshold for infection: A number of bacterial cells, greater than one, was needed before infection could occur, and anything less than this number would fail to saturate the host's defenses and infect. All the evidence is against Trevan's hypothesis; and the experimental data show that a single bacterial cell on its own can initiate infection.

Infection by single cells has been demonstrated experimentally by Hildebrand (1937, 1942) for two species of bacteria. He transferred single cells of *Erwinia amylovora* to the nectaries of flowers of apple trees. Many of the flowers became infected. He also transferred single cells of *Agrobacterium radiobacter** to tomato plants, and caused galls to form on them. Hildebrand's evidence about *A. radiobacter* has been confirmed recently by Manigault and Beaud (1967). Thyr (1968), working with *Corynebacterium michiganense,* found that only one bacterial cell was needed to initiate infection; and Goto (1972) found that a single cell of *Xanthomonas citri* injected into the mesophyll of citrus leaves can multiply and cause a leaf spot.

Evidence from disease/inoculum curves confirm that a single bacterial cell can infect, i.e., that there is no obligate synergism. The curves are those illustrated as Curves *A* or *B* in Fig. 1.7. Lippincott and Heberlein (1965a, b) found a linear relation between the number of *A. radiobacter* cells and the number of tumors they produced in Pinto bean leaves, and concluded that one bacterial cell is enough to initiate infection. Ercolani (1967), using the log dose/probit response curve (see Section 2.8), concluded that a single cell of *Corynebacterium michiganense* can infect tomatoes. In Fig. 1.18 his data are reproduced more simply. The estimated mean number of infections is plotted against the number of cells in the inoculum. For the two pure tomato cultivars, Roma and Fiorentina, disease is essentially proportional to inoculum, as in Curve *A* of Fig. 1.7, but in the line C1402 the curve turns to the right as in Curve *B* of Fig. 1.7, but more strongly. (For a detailed discussion see Section 2.2.) There is no indication in any of the curves of an upward curvature that would indicate synergism. Ercolani (1973), again using the log dose/probit response technique,

---

* Readers may be more familiar with the name *A. tumefaciens.* The nomenclature of *Agrobacterium* isolates is discussed by Keane *et al.* (1970).

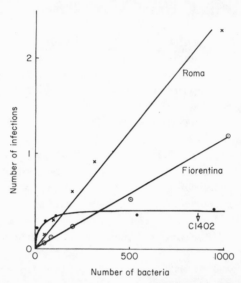

Fig. 1.18 The relation between the estimated number of infections of tomato plants and the number of cells of *Corynebacterium michiganense* in the inoculum. The number of infections is estimated as *m* in Eq. (1.1). (Data of Ercolani, 1967.)

extended his investigations to include *Pseudomonas lachrymans* on cucumber, *P. morsprunorum* on cherry, *P. phaseolicola* on bean, *P. syringae* on pear, *P. tabaci* on tobacco, and *P. tomato* on tomato. In all these combinations results were consistent with infection by a single bacterial cell. The data of Crosse *et al.* (1972) on the response of apple leaves to graded doses of *Erwinia amylovora* applied to the severed ends of the main veins, those of Pérombelon (1972) on the rotting of discs of potato tuber tissue inoculated with graded doses of *E. carotovora* var. *atroseptica* and those of Boelema (1973) with pea plants and *P. pisi* are also consistent with infection by a single bacterial cell. The evidence has been reviewed by Boelema (1973).

In another type of experiment Ercolani (1967) gave further evidence of infection by single cells. He inoculated tomatoes with a mixture of terramycin-susceptible and terramycin-tolerant variants of *C. michiganense* in approximately equal parts. When the dose of inoculum was weak (less than 1 ED_{50}), plants which became infected were found almost always to be infected with one variant or the other, but rarely with both. This indicates that infection started from a single cell. This result was confirmed by Ercolani (1973) in similar experiments with weak doses of the six *Pseudomonas* species in the susceptible hosts already discussed.

1.14   DISPUTED EVIDENCE FOR OBLIGATE SYNERGISM BETWEEN BACTERIA IN INCOMPATIBLE HOST–PATHOGEN COMBINATIONS

Ercolani (1973) claimed that there is a clear distinction between disease/inoculum curves of bacteria infecting compatible (homologous) hosts and those infecting incompatible (heterologous) hosts. In compatible combinations, one bacterial cell can start the infection process; in incompatible combinations, many bacterial cells are needed, i.e., there is a numerical threshold of infection. The distinction, Ercolani believes, is fundamental. With similar doses of inoculum, interaction between inoculated bacteria is more likely to occur when the host–pathogen combination is incompatible than when it is compatible, i.e., the host will dictate whether the bacteria interact. So, too, one may determine whether host–pathogen combinations are compatible or incompatible simply by determining whether the challenging bacteria are acting independently or cooperating during growth *in vivo*.

Although incompatible host–pathogen interactions are in effect immunity reactions and therefore strictly outside the scope of this particular chapter, the substance of Ercolani's conclusions contradicts the general thesis of what we have discussed. But there is good reason to dispute his conclusions.

Ercolani's evidence is of two sorts. The log dose/probit response curves for incompatible combinations differed sharply from those for compatible combinations. The incompatible combinations were *Pseudomonas lachrymans* on tobacco; *P. morsprunorum* on pear; *P. phaseolicola* on tomato, cucumber and tobacco; *P. syringae* on cherry, bean, cucumber, tomato, and tobacco; *P. tabaci* on cucumber, bean, and tomato; and *P. tomato* on bean, cucumber, and tobacco. The other sort of evidence came from inoculating incompatible host plants with mixtures of antibiotic-tolerant and antibiotic-resistant variants of the pathogen. The resulting hypersensitivity lesions were then found to contain both variants.

The evidence of Logan (1960), Klement and Goodman (1967a), and Hildebrand and Riddle (1971) indicates that hypersensitivity symptoms in incompatible combinations are observed only if the plant is inoculated with heavy doses of bacteria. The work of Klement and Goodman (1967b) and Klement (1971) shows that hypersensitivity symptoms appear after a comparatively short period even if the bacteria are killed only 2 or 3 hours after inoculation. Goodman *et al.* (1967) estimate that approximately 40 host cells must become necrotic before a lesion can be seen. Klement *et al.* (1964) found that after a necrotic lesion had developed in an incompatible host plant there was no further increase in lesion size.

The evidence Ercolani used for infection in incompatible host plants was the sudden appearance of dry necrotic areas, suggestive of a hypersensitive plant reaction. Examinations were apparently macroscopic, and this is where the trouble lay. Klement *et al.* (1964), also using the incompatible host–pathogen combination of *P. syringae* in tobacco, found that the number of lesions in tobacco was proportional to the concentration of bacterial cells in the inoculum, when they searched for lesions using a 10–12× magnification. That is, single bacterial cells on their own can infect, but the evidence for this requires the use of a lens or microscope, which it seems Ercolani did not use.

We reach several conclusions. First, in incompatible host–pathogen combinations, as in compatible combinations, a single bacterial cell can infect, and there is no evidence for obligate synergism. Second, a single cell of *P. syringae* in tobacco can multiply until its progeny causes a lesion visible under 10–12× magnification. Third, the multiplication of progeny does not always continue long enough to cause lesions visible to the naked eye; the host plant's immunity processes intervene and stop the multiplication before 40 or more host cells are killed. This conclusion holds for all six *Pseudomonas* spp. studied by Ercolani. Fourth, when one investigates whether single bacterial cells can infect an incompatible host plant, one must use an appropriate technique of scanning. For safety's sake, in case the host's immunity processes are strong, a microscope should be used that can detect an incompatible reaction within a single host cell. The difference between compatible host–pathogen combinations and incompatible combinations is the difference between susceptibility and immunity. If one wants to compare disease/inoculum relations in compatible combinations with those in incompatible combinations, one must necessarily allow for the intervention of immunity processes. (Immunity is defined and discussed in Chapter 7. We need not anticipate the discussion here, except to remark that in the terminology we shall use tobacco is immune from *P. syringae,* tomato from *P. phaseolicola,* and so on.)

1.15   THE COMPLEX VIRUS STORY. TOBACCO MOSAIC VIRUS.
INFECTION WITHOUT OBLIGATE SYNERGISM

Carsner and Lackey (1929) and Giddings (1946), working with the curly-top virus of sugar beets, put forward a theory of mass action between virus particles. They coupled this with the concept of a minimum dose of virus needed before infection of the host plants can occur, the equivalent for viruses of the concept of a numerical threshold for infection by fungi. The frequency with which leafhopper vectors transmit curly-top virus to healthy beet plants and infect them increases with the length of time the leafhoppers have previously fed on infected plants. This is indeed evidence that the chance of infection increases with the amount of virus in the vectors. But it is not evidence that more than one virus particle is needed; and one may reject as unproven their theory of a minimum dose for infection, in so far as it implies obligate synergism between virus particles of the same kind. In this respect viruses are probably not different from fungi or bacteria. But in recent results with many other viruses there is conclusive evidence for obligate synergism between virus particles of different kinds, i.e., between qualitatively different particles.

The virus story is complex. On present evidence, particles of tobacco mosaic virus are genetically complete, and a single particle on its own can initiate infection; the disease/inoculum relations of tobacco mosaic virus resemble those of fungi and bacteria. The particles of alfalfa and cowpea mosaic viruses are genetically incomplete, and for infection to occur there must be synergism between particles of different kinds, the particles supplementing one another genetically. Particles of a satellite virus are incomplete, and need the help of another virus for their replication; there is no proof that the "helper" virus benefits from its association with its satellite, and one cannot talk of synergism. The particles of potato spindle tuber virus are incomplete, and, on present evidence, seem to use their host plant's synthetic pathways to survive. These different patterns among the viruses are reflected in different patterns of disease/inoculum curves. In this section we consider the curve of tobacco mosaic virus.

There is a mass of data in the literature about the disease/inoculum relations of tobacco mosaic virus. But most of them are about artifacts irrelevant to natural infection in the field. Solutions of the virus have been rubbed on to leaves of *Nicotiana glutinosa* or *Phaseolus vulgaris,* and the local lesions that develop have been counted. In most of the experiments the purpose has been to devise techniques for assaying the

relative concentrations of virus in plant extracts or in solution. They
deal almost exclusively with an abundance of infections unknown in nature. For example, 10 or more lesions per cm^2 of leaf are common in
assays, whereas an average of only one infection per whole tobacco
plant would mean higher than average infection in the field. (If infections were randomly distributed, an average of one infection per plant
would mean that 63% of the plants in the field were diseased.) The essence of the artificiality is this. In local lesion assays a high proportion
of the available susceptible sites on the leaf is infected. In natural systemic tobacco mosaic disease only a very small proportion is infected.
When a high proportion of susceptible sites is infected, the shape of the
disease/inoculum curve is affected both by the total number of susceptible sites available and by differences in susceptbility between sites. See
Section 2.2. But neither of these factors materially affects the shape
when infection is low. In other words, when one uses experimental procedures that cause many lesions to develop per unit area of leaf, one
introduced complications highly relevant to techniques of virus assay
but almost wholly irrelevant to our present topic.

To avoid misunderstanding let it be stressed that the objection is to
the irrelevance of data and not their artificiality. Artifacts are necessary;
without them science would not advance far. But one can legitimately
try to avoid using data that because of their artificiality are clouded by
irrelevance.

Figures 1.19 and 1.20 show the effect of virus concentration on the
disease/inoculum relation. Figure 1.19, based on data of Kleczkowski
(1950) for tobacco mosaic virus on *Nicotiana glutinosa,* is representa-
tive of many experiments by various workers. It includes relatively very
high lesion numbers, such as are counted in the course of assays of
virus concentration. In contrast with Fig. 1.19, Fig. 1.20 selects data of
Furumoto and Mickey (1970) for low lesion numbers that would tally
better with numbers of systemic infections in the field. The disease/ino-
culum relation in Fig. 1.20 is that of Curve *A* and in Fig. 1.19 that of
Curve *B* of Fig. 1.7. The linear relation in Fig. 1.20 is supported by
the massed data of various workers. Scattered through the results of
Holmes (1929), Caldwell (1933), Samuel and Bald (1933), Chester
(1934), Youden *et al.* (1935), and Kleczkowski (1950) are occasional data for tobacco mosaic virus concentrations low enough to be
considered relevant. The criterion for relevance was that the ratio Y/N
should be less than 0.05, these symbols having the meanings given in
Section 1.12. These data, pooled, fully confirm those of Furumoto and
Mickey: At low concentrations of virus the disease/inoculum curve is

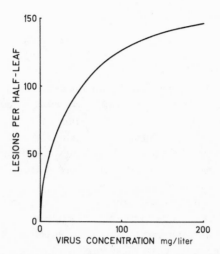

Fig. 1.19 The relation between the number of lesions per half-leaf of *Nicotiana glutinosa* and the concentration of tobacco mosaic virus, in milligrams per liter, in the inoculum. (Data of Kleczkowski, 1950.) Judged by lesion numbers, only a few microns of this curve near the origin is relevant to natural infection; the rest is artifact.

practically a straight line from the origin. Disease is proportional to inoculum; and infection starts from a single virus particle acting independently of all others. There is no obligate synergism in the infection process.

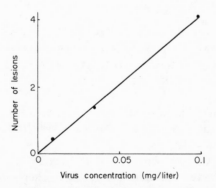

Fig. 1.20 The relation between the number of lesions per half-leaf of *Nicotiana glutinosa* and the concentration of tobacco mosaic virus, in milligrams per liter, in the inoculum. (Data of Furumoto and Mickey, 1970, for the lowest three concentrations of virus strain *U*1.) The linear relation here is less artificial than the curvilinear relation in Fig. 1.19; it seems to hold generally for data by workers on tobacco mosaic virus, but does not hold for all viruses.

Although there is no obligate synergism, there is evidence for facultative synergism, even if only at artificially high concentrations of virus. Hulett and Loring (1965) found that small particles and standard particles of tobacco mosaic virus interacted synergistically at high concentrations. Here we have an analogy with *Puccinia graminis*. There is no obligate synergism; single spores of *P. graminis* infect on their own when the spore concentration is low; but spores interact with facultative synergism when their concentration is high. (See Section 1.5.) The chemical reasons for the facultative synergism is fungi differ entirely from those in viruses. (It will be remembered from Section 1.5 that uredospores of *P. graminis* release higher straight-chain aldehydes and alcohols which stimulate germination.) But in the present topic of disease/inoculum curves, tobacco mosaic virus falls in line with fungi.

The conclusion that tobacco mosaic virus infects without obligate synergism is for the intact virus. The infectious nucleic acid of tobacco mosaic virus may behave differently, and Bawden and Pirie (1959) have described an abnormally steep dilution curve for it. This seems artificial and irrelevant to the present topic.

Most data on infection by viruses other than tobacco mosaic virus are too few at relevant low concentrations for us to judge about obligate synergism. On the data of Price (1938) and Kleczkowski (1950) for infection of *Vigna sinensis* and *Phaseolus vulgaris,* at least some strains of the tobacco necrosis virus complex seem to behave like tobacco mosaic virus and infect without obligate synergism.

1.16   SOUR CHERRY RINGSPOT, PRUNE DWARF, COWPEA MOSAIC, ALFALFA MOSAIC, AND TOBACCO RATTLE VIRUSES. OBLIGATE SYNERGISM BETWEEN QUALITATIVELY DIFFERENT PARTICLES OF THE SAME VIRUS

There is evidence that infection by these viruses involves obligate synergism, and that this synergism requires the virus to have two or more different sorts of particle.

Figure 1.21 reproduces data of Fulton (1962) in the form of a disease/inoculum curve. The data are for sour cherry necrotic ringspot virus in sap inoculated onto leaves of *Momordica balsamina*. Virus concentration is recorded on an arbitrary scale with undiluted sap taken as 100. Synergism is apparent. The curve is nearly what one would expect in a Poisson curve if at least three virus particles had to interact syner-

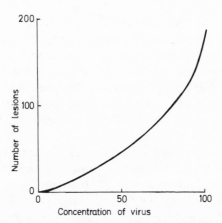

Fig. 1.21   The relation between the number of lesions on *Momordica balsamina* and the concentration of sour cherry necrotic ringspot virus. The concentration of undiluted sap is taken as 100. The curve is nearly what one would expect if at least three virus particles had to interact synergistically to cause infection. (Data of Fulton, 1962.)

gistically to cause infection; i.e., the curve is approximately a three-hit curve.

Fulton found that the host plant affected the slope of the disease/inoculum curve. The curve was close to a three-hit curve when sour cherry necrotic ringspot virus was tested on *M. balsamina,* and close to a two-hit curve when the same virus was tested on *Dolichos biflorus.* One might perhaps suggest that sour cherry ringspot virus needs a more complete genome to cause local lesions on *M. balsamina* than *D. biflorus.* There is nothing inherently improbable about such a suggestion.

Working with another virus, prune dwarf virus, on *Sesbania exaltata* Fulton also obtained a disease/inoculum curve close to a three-hit curve.

However, a very different interpretation can be put on Fulton's results. It seems that he counted macroscopic lesions. It is possible that single particles did infect, but that the infection of a single cell did not easily cause the collapse of the 40-odd neighboring cells which must die if a lesion is to be visible to the naked eye, and that the process of collapse following infection by sour cherry necrotic ringspot virus stopped earlier in *M. balsamina* than in *D. biflorus.* The following rule is necessary in interpreting the slopes of disease/inoculum curves obtained by counting local lesions: If the slope is consistent with infection by single particles, that is, if the curve is a one-hit curve, the evidence of the slope can be accepted even if only macroscopic lesions are counted; but if the slope indicates that two or more particles are needed for infection, the evidence can be accepted only if results are confirmed by scanning with a strong lens or a microscope. The rule recognizes that confluent lesions inevitably occur; that the Poisson curves confuse confluence with obligate synergism; and that counting only macroscopic lesions might sometimes introduce a bias toward counting confluent lesions. The other examples is this Section are free from the taint of possible confusion.

Cowpea mosaic virus was studied by Van Kammen (1968). The evidence is clear, first, that two particles must interact synergistically before infection can occur, and, second, that these two particles must be qualitatively different from one another. From plants infected with cowpea mosaic virus, Van Kammen isolated two components by centrifugation: a middle component of particles with 23% ribonucleic acid and a bottom component of particles with 32%. (The top component, without nucleic acid, is irrelevant to our story.) Neither the middle nor the bottom component, each on its own, was able to infect the test plants of *Phaseolus vulgaris*. Mixed, they infected readily. Figure 1.22 reproduces some of Van Kammen's results. It shows, on the top, the effect of increasing the concentration of middle component in a mixture with a constant concentration of bottom component, both components on their own being noninfectious. Reversing the procedure and adding increasing amounts of bottom component to a constant amount of middle component gave similar results, shown at the bottom of Fig. 1.22.

Obligate synergism during infection has been shown for alfalfa mosaic virus as well. Price and Spencer (1943) studied the effect of the concentration of this virus on the number of local lesions produced on leaves of *Phaseolus vulgaris*. The slope of the disease/inoculum curve suggested synergism, but unfortunately the details they published were not in a form that allows critical inquiry. The more recent work of Van Vloten-Doting *et al.* (1970) has demonstrated obligate synergism beyond doubt. By centrifugation these workers purified five nucleoprotein components, three of which concern us here: the bottom component, the middle component, and one of the top components. The particles differed characteristically, e.g., in length, from component to component. The particles from any one component were unable to cause infection; but when the three components were mixed, the mixture infected readily.

Tobacco rattle virus (Harrison and Woods, 1966; Sänger, 1968a,b; and Huttinga, 1972) is another virus with obligate synergism between qualitatively different particles.

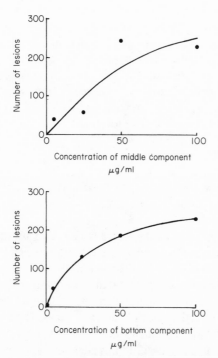

Fig. 1.22   *Top.* The relation between the number of local lesions produced by a solution containing 25 μg per milliliter of the bottom component of cowpea mosaic virus and the concentration of the middle component.

Bottom. The corresponding relation between the number of lesions produced by a solution of 25 μg of the middle component and the concentration of the bottom component. The host plant was *Phaseolus vulgaris*. (Data of Van Kammen, 1968.)

The current explanation is that the obligate synergism is genetic supplementation. The separate components of cowpea mosaic and alfalfa mosaic viruses cannot, each on its own, replicate themselves because the genetic material—the nucleic acid—is inadequate. But when the appropriate components are mixed, the mixture has all the information needed for replication. In other words the genetic information is spread over two or more particles of the multicomponent virus. The genome is divided.

Here, on the theory of genetic supplementation, we must see viruses in perspective against the background of pathogens in general. Compared with bacteria and fungi, viruses are small. The pathogen must introduce into the host plant a genome with all the information needed for its own replication. For viruses, the information must be adequate for the synthesis of all specific products that lead to the replication of nucleic acid and the formation of a specific protein coat (except in coatless viruses). Some virus particles fall below the threshold in size that can contain all this information; and the synergistic pooling of information from several particles is needed. Here we are dealing with a need not known to occur in bacteria and fungi with their relatively large, organized nuclei and chromosomes.

1.17   INCOMPLETE VIRUS PARTICLES THAT INFECT WITHOUT OBLIGATE SYNERGISM. POTATO SPINDLE TUBER VIRUS. SATELLITE VIRUSES

Not all viruses with incomplete genomes live by obligate synergism. Potato spindle tuber virus is incomplete. Its molecular weight is as low as 60,000 daltons (Diener, 1971), which is too little for it to code for is own replication.* (We can put this weight in perspective by noting that the 300 mμ particle of tobacco mosaic virus—ordinarily regarded as the smallest infective particle of this virus—is 1000 times as heavy and contains 50 times as much nucleic acid as a particle of potato spindle tuber virus.) Yet, despite this deficiency, the disease/inoculum curve is for infection by single particles (Raymer and Diener, 1969). It is a one-hit curve.

Parasitism by spindle tuber virus seems to be the most intimate conceivable form of obligate parasitism. But there is no obligate synergism between particles. It seems possible that spindle tuber ribonucleic acid is accepted by the host and replicated by the host's synthetic pathways. That implies intimate obligate parasitism, but not synergism. Virus particles are helped by the host, not by the other particles.

Satellitism, discovered by Kassanis and Nixon (1961), is another phenomenon in which incomplete virus particles do not imply synergism. The satellite virus of tobacco necrosis virus cannot infect on its own, but becomes infectious in the presence of tobacco necrosis virus.

---

* There have recently been further estimates of the molecular weights of the related pathogenic ribonucleic acids from citrus exocortis disease (Semancik *et al.,* 1973) and potato spindle tuber disease (Sogo *et al.,* 1973). The estimates vary with the experimental method—electrophoretic mobility, inactivation by ionizing radiation, and visualization by electron microscopy—but agree in establishing that the weights are very low.

So too the satellite virus of tobacco ringspot virus infects only in the presence of tobacco ringspot virus. Apart from this restriction the satellite viruses are no weaklings. In infected plants the satellite viruses of tobacco necrosis and tobacco ringspot viruses become more abundant than tobacco necrosis and tobacco ringspot viruses themselves (Schneider, 1971).

Genetic inadequacy is the current explanation of satellitism. The satellites are too small to reproduce on their own and depend on helper viruses for supplementary nucleic acid. Thus, Schneider (1971) found that the strands of nucleic acid of the satellite virus of tobacco ringspot virus have a molecular weight of only 86,000 daltons.

The sort of disease/inoculum curve in satellitism has not been determined but varies presumably with the circumstances. Single particles of satellite virus could presumably infect a plant already infected with the helper virus. But the simultaneous infection of a healthy plant with both helper and satellite viruses would require at least two particles to be present at a susceptible site, and this would be reflected in the disease/inoculum curve.

1.18   RESTRICTIONS IMPOSED ON THE PATHOGEN BY OBLIGATE SYNERGISM. VECTOR TRANSMISSION. POSSIBLE BENEFITS FROM OBLIGATE SYNERGISM

Obligate synergism restricts transmission, and probably requires the dispersal of particles to be nonrandom.

Consider a calculation. Suppose that one were transmitting a virus by rubbing a solution of it randomly on susceptible leaves. Suppose the leaves of the plant had a million susceptible sites. This, to judge by the estimates of Kleczkowski (1950) for various viruses, would be a realistic if somewhat conservative figure to use for a well-grown plant; in any case, the actual figure we use does not materially affect the gist of the calculation, and we can proceed without further ado about it. Suppose finally that we were using a virus in which two different sorts of particle were needed for infection, i.e., suppose that the infection process involved obligate synergism between two qualitatively different particles. Then to get these two qualitatively different particles together at a susceptible site we would need a solution at least 2000 times as concentrated as would be needed to get a single particle at the site. That is, other things being equal, with obligate synergism as compared with no synergism, the virus solution would need to be at least 2000 times as strong to produce the same amount of disease. The estimate, 2000 times, is a minimum, and applies to solutions containing equal amounts of the two sorts of particles. If the amounts of the two sorts were unequal, the solution would have had to be more than 2000 times as strong. We are concerned in this calculation with the occupation of only one susceptible site per plant, because to cause systemic disease it is necessary only that one susceptible site per plant be infected. That is, the calculation is for systemic disease, as it ought to be for virus diseases.

Here are details of the calculation, using the Poisson equations in Section 1.12. By hypothesis, $N = 1,000,000$ and $Y/N = 10^{-6}$. By the equations, $Y/N = 10^{-6}$ if $ax = 10^{-6}$, for at least one particle at the susceptible site, or $ax = 10^{-3}$, for at least two particles at the site. Other things than the virus concentration x being equal, a has the same value throughout. Therefore, to get two randomly distributed particles at the infection site, x would have to be 1000 times as great as to get only one particle at the site. But the Poisson equations are concerned only with particles that are alike, i.e., of one sort. So to get two randomly distributed particles of different sorts at the susceptible site, the virus concentration would have to be 2000 times as great as to get only one particle at the site, if the two sorts were equally abundant, 2250 times as great if the two sorts were present in the ratio 1:2, 3125 times as great if they were present in the ratio 1:4, and so on.

If three different sorts of particles must interact synergistically, the efficiency of random transmission is even more greatly reduced.

To return to our calculations, N is necessarily large. To say that the concentration of virus required to cause one infection is at least 2000 times as great when two different particles are needed as when only one particle is needed is therefore almost exactly the same as to say that for a given concentration of virus the number of infections would be at least 2000 times as great when only one particle is needed as when two different particles are needed. Obligate synergism between two different and randomly distributed particles would reduce the efficiency of virus transmission so much that in a field or orchard it would be as easy for at least 2000 infections with one particle to occur as for one infection to occur with two different particles. Consider this reduced efficiency in relation, first, to infection of clones or long-lived trees, and, second, to virus transmission by vectors in annual crops.

Inefficient transmission is compatible with survival, provided that rare transmission is compatible with survival. Rare transmission is compatible with survival if the host plants are long-lived clones or trees, because rarity of transmission must be judged against the longevity of the host plants. It need cause no surprise that obligate synergism has been found in viruses inhabiting fruit trees. This could be true even if they had no very efficient vectors.

The essence of obligate synergism between particles is that particles of different sorts must come together. Vectors bring them together. The key to our calculations has been the random distribution of particles; and the results of the calculations emphasize that with random distribution the coming together of particles is relatively rare. Transmission by vectors breaks the difficulty of obligate synergism by breaking the randomness of the distribution. In vector transmission, particles of all sorts are brought together and transmitted to a site; and the chance of the different particles coming together in vector transmission is far greater than when their distribution is random.

To survive in annual crops a virus must be frequently transmitted. In annual crops it seems likely that obligate synergism between particles makes the virus greatly dependent on transmission by vectors. Possession of a vector brings an advantage to any virus; it brings a double advantage to viruses when the genome is divided between two or more particles.

Cowpea mosaic and alfalfa mosaic viruses are transmitted by insects, tobacco rattle virus and the satellite virus of tobacco ringspot virus by nematodes, and the satellite virus of tobacco necrosis virus by a fungus.

There may be benefits to the virus from obligate synergism to counterbalance the restrictions it imposes. Obligate synergism is associated with genetically different particles. The ribonucleic acid viruses have a low recombination frequency; and Cooper (1968) suggested that the exchange of genetic material would be facilitated if several particles were needed to constitute the infectious virus. Jacobson and Baltimore (1968) suggested that protein synthesis might be facilitated if several particles were needed. We could add a third suggestion, based on the theory that the simpler the genome of an obligate parasite, the greater is its host range. See Section 7.21. The range includes vectors as hosts. On this theory the host's immunity processes tend to have more difficulty getting a grip on a simple genome than a complex one. Whereas the first two suggestions emphasize the divisibility of the infectious virus, the third emphasizes the genetic simplicity of its parts. All three suggestions could, of course, have merit. But one must beware of trying to see a virtue (for the virus!) in what may be only a necessity. One must not assume *a priori* that obligate synergism is a beneficial adaptation for the virus simply because it occurs. Obligate synergism could also be an unavoidable weakness arising from an incomplete organization of the virus, a weakness that has not prevented the survival of the virus largely because of an environment of vectors.

1.19   DISEASE/INOCULUM CURVES IN RELATION TO TRANSMISSION BY VECTORS

There is evidence that the dose of virus transmitted by vectors varies in size. Many years ago Carsner and Lackey (1929) and Giddings (1946) showed that the number of transmissions of sugar beet curly-top virus increased with the time the leafhopper vectors fed on a source plant of the virus. One infers from this that the available dose is usually more than the minimum needed for infection.

With the same dose per vector, what is the relation between the number of transmissions and the number of vectors? That is the question probed in this section.

Hoggan (1933) used the aphid *Myzus persicae* to transmit cucumber mosaic virus from tobacco to tobacco, and found that the percentage of infection increased with the number of aphids. Watson (1936), also using tobacco and *M. persicae,* transmitted the virus *Hy* III with varying numbers of vectors and varying feeding times on the healthy plant. Storey (1938) transmitted maize streak virus with varying numbers of the leafhopper vector *Cicadulina mbila;* and Posnette and Robertson (1950) transmitted *Theobroma* virus 1A and 1B to cacao embryos with varying numbers of the mealybug vector *Pseudococcus njalensis.*

Figure 1.23 illustrates the results of Posnette and Roberson. The points are averages for *Theobroma* viruses 1A and 1B, which behaved similarly. A straight line is fitted by standard procedures to the points. It passes close to the origin, the departure being insignificant. The relation between disease and the number of vectors is the disease/inoculum relation of Curve A of Fig. 1.7. Disease is proportional to inoculum, the inoculum coming in the form of doses of virus injected by the insect vectors.

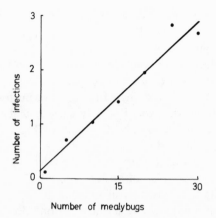

Number of mealybugs

Fig. 1.23   The relation between the estimated number of infections in cacao embryos by *Theobroma* viruses 1*A* and 1*B* and the number of mealybug vectors per embryo. The number of infections was estimated as *m* in Eq. (1.1). (Data of Posnette and Robertson, 1950.)

Figure 1.24 illustrates the results of Watson. The curve is for a 3-hour feeding period; the curves for other feeding periods are similar. The relation between disease and the number of vectors is the disease inoculum relation of Curve *B* of Fig. 1.7. Presumably all the test plants

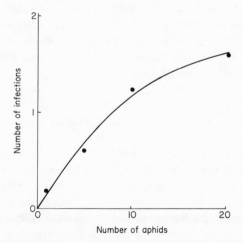

Number of aphids

Fig. 1.24   The relation between the estimated number of infections in tobacco plants by virus *Hy* III and the number of aphid vectors per plant. The number of infections was estimated as *m* in Eq. (1.1). (Data of Watson, 1936.)

were not identical in susceptibility. Storey's results with maize streak virus also conform with Curve B.

The story of fungus vectors seems to be much the same as that of insect vectors. Teakle (1973) transmitted tobacco necrosis virus to roots of *Phaseolus aureus* and *Vigna sinensis* by means of zoospores of *Olpidium*. The experiments were designed for another purpose and the data are consequently rather meager. But they suggest that the disease/inoculum curve with *P. aureus* is like Curve A of Fig.1.7 and the curve with *V. sinensis* like Curve B.

Curves A and B in Fig. 1.7 are for independent action. With vector transmission the injected dose of virus is the infectious entity, and the experimental results are consistent in showing that the doses behave independently of one another. The particles in one dose do not interact with the particles in any other dose.

This bears on the story of obligate synergism: Obligate synergism does not affect the shape of disease/inoculum curves if the virus is vector transmitted. If particles in one dose are independent of particles in another dose, there cannot be synergism between doses; and if there is no synergism between doses, the disease/inoculum curves cannot reflect synergism, because the curves reflect only interaction between doses. Synergism within a dose makes the dose infective. But only synergism between doses could be reflected in upward curvature of the disease/inoculum curve. On the evidence available, the disease/inoculum curves for cowpea or alfalfa mosaic virus, or any other virus with a divided genome, when the virus is vector transmitted, would have the same form as curves for fungi, bacteria, or viruses such as tobacco mosaic virus that infect without obligate synergism.

1.20 INFECTIOUS ENTITIES

Rapilly (1968), Rapilly and Fournet (1968), Rapilly *et al.* (1970), and Stanbridge and Gay (1969) introduced the concept of units of dispersal. See Section 1.5. Uredospores of *Puccinia striiformis* are dispersed in clumps, a mucilaginous coating holding them together. A clump and not an individual uredospore within the clump is the unit of dispersal. So too when a vector injects a dose of virus into a host plant, the whole dose and not an individual virus particle is the unit of dispersal.

We must introduce a corresponding concept of a self-contained unit of infection, or infectious entity. The uredospores of *P. striiformis* in a mucilaginous clump infect as an entity. Only one uredosorus starts from a clump, independently of how many spores in the clump germinate and infect. The virus particles injected by a vector as a dose at one site infect as an entity. Only one systemic infection starts from them. This is obvious, because systemic disease can be defined as disease in which a single act of infection is enough to infect the whole susceptible part of the plant. Single fungus spores, bacterial cells, or virus particles are also infectious entities, when they are able to infect on their own.

Webster's dictionary defines an entity as something that has a unitary and self-contained character, and an independent, separate, or self-contained existence. An infectious entity, as we define the term, is the actual entity that brings about infection, be this infectious entity a single fungus spore or bacterial cell or a relatively massive dose of virus particles carried in a vector.

Synergism is sometimes involved within a complex infectious entity. Within the mucilaginous clump of uredospores of *P. striiformis* the spores help one another to infect, as Rapilly *et al.* (1970) showed. This synergism is facultative, of the sort mentioned in Section 1.5, and probably the result of the mutual stimulation of germination by substances released from the spores. Obligate synergism between incomplete and qualitatively different virus particles is also relevant and made possible when the particles coexist as an entity within a vector.

Units of dispersal and infectious entities are separate concepts. To refer to an artifact for illustration's sake, if randomly scattered particles in a solution of cowpea mosaic virus are rubbed on a leaf of *Phaseolus vulgaris* the random individual particles are the units of dispersal but the random pairs of complementary particles are the infectious entities. In this illustration, units of dispersal may be 2000 times more common than infectious entities (see Section 1.18). But, outside of artifacts, units of dispersal and infectious entities are probably often similar or identical.

We choose the word entity in preference to unit, because entity carries the meaning of being self-contained. Thus, we write of units of dispersal, because self-containment is not implied, but of infectious entities, because self-containment is implied. Self-containment is an inherent, internal quality. It does not refer to the environment. Thus, a uredospore of *Puccinia graminis* is an infectious entity (provided it is not defective), but infection requires that the spore should be on a susceptible host plant and that the temperature, humidity, etc., should permit infection. There are necessarily many more infectious entities than acts of infection.

Complex infectious entities behave like simple entities in disease/inoculum curves. It seems to matter little, or not at all, whether the infectious entity is a single fungus spore, bacterial cell, or virus particle or a complex aggregation of spores, cells, or particles. Curves *A* and *B* of Fig. 1.7—the curves for independent action—are found for vector doses of virus as well as for single spores, cells, or particles. Experimental evidence was given in the preceding section. There is no apparent reason why Curve *C,* for antagonistic interaction between infectious entities, or Curve *D,* for synergistic interaction, should not be found for complex entities. The synergism here can only be facultative; obligate synergism is necessarily absent between infectious entities, be they simple or complex, because by definition an entity is self-contained.

1.21   A Basic Principle of Infection: The Probabilistic Viewpoint

The principle is this: Two or more fungus spores, bacterial cells, or virus particles cannot infect if one of the same sort cannot infect. Stated differently, there is no obligate synergism between spores, cells, or particles of the same sort.

The key words here are "of the same sort." The principle does not relate to different sorts of spores, cells, or particles. For clarity, two ex-· amples are given of combinations of different sorts of spores, cells, or particles, in order to exclude them from further discussion about the principle. The first example was discussed in Section 1.16. Single particles of cowpea or alfalfa mosaic virus cannot infect, but an appropriate combination of particles of different sorts can. Each particle is genetically deficient; an adequate genome for infection is provided only by combining two particles differing from each other. This can be regarded as sexuality of the most primitive kind. Combinations of this kind seem to be confined to viruses. The second example concerns sexuality as it occurs commonly in fungi. Two sexes or mating types may supplement each other to give a genetic apparatus appropriate to infection. Here, too, different sorts are involved; and sexuality automatically falls outside the scope of the principle.

The principle, that two or more spores cannot infect if one spore of the same sort cannot, contradicts Gäumann's theory of a numerical threshold of infection. The evidence against this theory was discussed in detail in Section 1.10. What is relevant now is that Gäumann saw his theory in terms of spores of the same sort. Quantitative deficiency, he believed, could be countered by force of numbers as distinct from differences in sort. Thus, a single infection spot of *Synchytrium endobioticum* in potatoes cannot, he believed, develop into a wart because the morphogenic stimulus is too weak. This implies that *S. endobioticum* is genetically fit to infect, but that the infection does not proceed beyond an infection spot unless other spots of the same sort join in to help. But the disease/inoculum curve for wart disease contradicts equally both synergism between myxamoebae and synergism between resulting infection spots; and for the implication that quantitative deficiency can be corrected without qualitative variety there is no evidence at all.

It is convenient and realistic to think of infection in terms of probabilities. If it takes an average of a million particles of tobacco mosaic virus to infect a tobacco plant, the probability that any one particle will infect is one in a million. No active role is implied for the remainder. Buller (1931) estimated that some basidiomycetes produce 10^9 to 10^{12} spores per fruiting body; in stable populations of these fungi the probability that any one spore will infect is of the order of 10^{-12} to 10^{-9}. No active participation by the remainder is implied.

Ponder, if you will, small probabilities of infection and the waste of propagative matter; but ponder them against the background of universal prodigality and waste in reproduction. Wasted pollen lies thick on the floor of a pine forest. The female cod lays a million eggs, and only two of the progeny need survive and reproduce to maintain the population. Spermatozoa are produced by the million. In reproduction the probability of success is low; and waste is part of the process. Waste in infection follows the general pattern; and in high numbers of spores, cells, or particles there is no need always to look for numerical thresholds of infection or other active roles for all the many infectious entities that do not chance to infect and so go to waste.

1.22   DISEASE/INOCULUM RELATIONS NEAR THE ORIGIN: INDEPENDENT ACTION

It is the stated purpose of this chapter to examine the relation between the amount of disease and the amount of inoculum that produces it. It is the purpose of this section to summarize the evidence for relations near the origin, i.e., the evidence for small amounts of disease such as might be found during the early stages of an epidemic.

Near the origin, disease/inoculum curves follow two rules. *One,* the curve starts at the origin. *Two,* the curve is for all practical purposes a straight line. Evidence that the curve starts at the origin was fully discussed in Section 1.11. Evidence about its straightness must now be probed in more detail.

A straight line near the origin and starting from the origin means that disease is directly proportional to inoculum; or, conversely, evidence that disease is proportional to inoculum is evidence that the line is straight and starts from the origin.

Considered over the full range of disease and inoculum, there are four known disease/inoculum curves, illustrated by Curves *A, B, C,* and *D* in Fig. 1.7, which in turn repeat the curves in Figs. 1.1–1.4. Curves *C* and *D* in Fig. 1.7 are for antagonistic interaction and synergistic interaction, respectively. Interaction destroys direct proportionality. But all the evidence is that these interactions occur measurably only at high levels of inoculum. (It will be remembered here that the synergism concerned in Curve *D* is facultative.) Nearer the origin their influence is negligible, and they fall outside the scope of this section. In Curve *B* there is no antagonistic or synergistic interaction, but there is competition between infective entities for susceptible sites. But this competition is marked only at high concentrations of inoculum, and is necessarily unimportant at low concentrations; so it too falls outside the scope of this section. In Curve *A* there is neither antagonism, synergism, nor competition; and disease is proportional to inoculum over the whole range of data. Curve *A* has been found with fungus spores, bacterial cells, and viruses transmitted by vectors (see Fig. 1.23). Curve *A* is an abbreviated form of Curve *B,* being found when the number of infectious entities is small relative to the number of susceptible sites.

At low levels of inoculum and disease, when x and Y are small, the one-hit Poisson equation—Eq. (1.2)—discussed in Section 1.12 reduces approximately to

$$Y = aNx$$

Approximately, for all practical purposes, Y is proportional to x, when x is small, and the straight line relating them has the slope aN. The Poisson equation almost certainly oversimplifies disease/inoculum relations—a matter discussed in Chapter 2—and one should replace aN by a series

$$a_1N_1 + a_2N_2 + \cdots$$

But the effect remains the same: Disease is directly proportional to inoculum, when the amount of inoculum is small.

The concept of the disease/inoculum curve as a straight line immediately after leaving the origin can be regarded as an experimental law in the sense of the term given in Section 1.11. Alternatively it can be regarded as an hypothesis of independent action when units of dispersal of fungus spores, bacterial cells, or virus particles are too few and far between to interact or compete with one another. On either alternative, the only known contrary evidence is from artifacts, as when cowpea mosaic virus or other virus with a divided genome is transmitted mechanically instead of by natural vectors. Put differently, the only known contrary evidence is from units of dispersal which are not infectious entities; and the only known inherently noninfectious units of dispersal are artifacts.

On the practical side, the experimental results are reassuring. Plant pathologists are busy simulating disease epidemics in computer programs. In these programs it is always assumed that (with due allowance for multiple infection) disease is proportional to inoculum. The evidence shows that this assumption fits the known facts when the amounts of disease and inoculum are small, as they are during the early stages of an epidemic.

Chapter 2

More about Disease/Inoculum Curves

2.1 THE SCOPE OF THE CHAPTER

Like Chapter 1, this chapter is about disease/inoculum relations. The central questions are the same. If 1000 spores fall on plants and cause n lesions to develop, how many lesions would 2000 spores have caused? How does inoculum dose affect disease response?

Chapter 2 goes back to the four known disease/inoculum curves: Curves A, B, C, and D of Fig. 1.7, which represent the experimental data in Figs. 1.1–1.4. The next two sections deal with Curves A and B, which are the basic curves of disease-response/inoculum-dose relations, and suggest a model. Sections 2.4 and 2.5 discuss Curves C and D, which are probably largely artifacts, and Section 2.6 gives a short summary of the four curves.

The rest of the Chapter is concerned with miscellaneous topics. There is a section on disease/inoculum relations in root disease. There are arguments in the literature that disease/inoculum curves for root diseases are inherently different from those for foliage diseases because soil is a three-dimensional medium. These arguments are critically probed in Section 2.7. Section 2.8 discusses tests for the independent action of spores as inoculum. Section 2.9 questions the cult of using the logarithm of the amount of inoculum in graphs. Section 2.10 discusses the use and interpretation of parameters in disease equations.

2.2 THE TWO PATHS TO SUSCEPTIBILITY AND THEIR RELATION TO CURVES A AND B OF FIG. 1.7. A HITHERTO NEGLECTED FACTOR IN EPIDEMIOLOGY

Consider green mold of oranges caused by *Penicillium digitatum*. It is generally accepted, and for our purpose need not be doubted, that the fungus enters only through wounds. For a start we shall oversimplify the model and assume that the wounds are all alike and the spores are randomly distributed. Then we can apply the Poisson one-hit equation (1.2) given in Chapter 1 and reproduced here for convenience of reference

$$Y = N\,(1 - e^{-ax})$$

Here Y is the number of infections that occur in any uniform group of oranges (a box of oranges, or a ton, or a carload—it does not matter); x is the number of spores of *P. digitatum* in that group; N is the total number of wounds in the group; and a is a parameter reflecting the susceptibility of the wounds to infection.

In the equation Y, the disease response, is related to x, the spore dose, by two parameters, N and a. These two parameters represent the two paths to susceptibility. First, a group of oranges is more susceptible to mold if N is great, i.e., if there are many wounds. For this reason packhouse managers try to reduce wounding; care is taken when the fruit is picked to avoid clipper cuts; gravel or other abrasive material is cleaned out of the field boxes; machinery is examined for projections that could pierce the rind; packers wear gloves to prevent fingernail cuts; and so on. Second, a group of oranges is more susceptible if a is great, i.e., wounds are easily infectible. Thus, for example, low temperatures reduce a, and for that reason fruit is refrigerated. (And of course packhouse managers try to reduce the spore load x by sanitation or fungicides; but for the purpose of this chapter, x is the independent variable and not a parameter.)

These two paths to susceptibility differently affect the disease/inoculum curve. Great susceptibility via a high value of N, i.e., because there are many wounds, reduces curvature; great susceptibility via a high value of a, i.e., because wounds are easily infectible increases curvature. With high N but low a, the disease/inoculum curve tends toward Curve A of Fig. 1.7 (i.e., a straight line). With low N but high a it takes the form of Curve B (i.e., curving to the right).

We formulate a general principle of infection: Susceptible sites vary both in quantity, i.e., in number, and in quality, i.e., in infectibility; these variations determine the shape of the disease/inoculum curve and hence the course of an epidemic. This principle is the topic of this and the next section. We need not confine ourselves to simple wound parasites or assume the simple Poisson equation. We take it as axiomatic that, whatever the disease may be, susceptible sites may vary both in quantity and quality. What remains is to suggest a realistic model that will indicate the consequences of these variations.

A susceptible site is where the pathogen can enter the host. It may be a wound or a natural opening such as a stoma, or where a fungus can penetrate the intact surface. There is evidence that even when the intact surface can be penetrated the number of susceptible sites is often limited. The pathogen and the environment are as much involved in the quantity and quality of the susceptible sites as the host. A susceptible site is where the pathogen can enter the host if the environment permits it to do so; and parameters describing the quantity and quality of susceptible sites are inevitably determined by the whole disease triangle of host, pathogen, and environment.

The notion of competition needs probing. When the disease/inoculum curve is straight, as in Curve A of Fig. 1.7, we infer that only one spore infects per wound. In effect, the spore has the wound to itself. But when the curvature starts, as in Curve B, we infer that some wounds have two or more germinating spores, all in a position to infect. We think then of competition between them. It does not matter whether one spore germinates first and monopolizes the attack, or whether two or more spores germinate more or less simultaneously so that the mycelium in the infection has several origins. The effect is competition. We can think of it under different names. We can think of it as a superfluity of spores in a wound or of multiple infection (if this occurs). Competition, as we use the word here, means that there are more spores available at the wound than are needed for the job. No antagonism or synergism is implied. Competition has the same meaning here as when there is competition between two or more applicants competing for appointment to a chair of professor; there is a superfluity of applicants.

There is a simple mathematical approximation applicable to the Poisson equation. When ax is very small, e^{-ax} is numerically very nearly equal to $1 - ax$, and hence $1 - e^{-ax}$ is very nearly equal to ax. Thus, mathematical tables show that if $ax = 0.001$, then $e^{-ax} = 0.999000$, so that $1 - e^{-ax} = 0.001000$; that is, the approximation is correct to six decimal places. So too when $ax = 0.01$, then $1 - e^{-ax} = 0.00995$; and the approximation has an error of 1 in 200, which usually is small enough to ignore.

When N is great so that Y/N is very small, ax is also very small. The approximation then reduces the Poisson equation to

$$Y = Nax \qquad (2.1)$$

This is the equation for a straight line passing through the origin, i.e., for Curve A. When the Poisson equation applies, a straight disease/inoculum line through the origin means that N is great relative to Y, i.e., there are relatively many uninfected susceptible sites still available for infection. In other words, there has been insignificantly little competition between spores for susceptible sites. By the same token, curvature implies competition for sites; and when the disease/inoculum line curves as in Curve B, this means that Y/N and ax are no longer small, and that the simplifying approximation behind Eq. (2.1) no longer applies.

If circumstances were to make the Poisson equation valid, the equation would represent both Curves A and B, depending on the numerical value of the parameters. But the equation is not valid. In terms of the model of green mold of oranges, the equation requires all wounds to be alike. But they are not all alike. A deep clipper cut that allows juice to ooze over the rind is far more liable to infection than a superficial abrasion that dries quickly.

So let us divide wounds into classes according to the danger of their being infected, i.e., according to the probability of infection by a spore. Of wounds in, say, danger class 1 there are N_1, and the parameter defining their susceptibility is a_1. Of wounds in class 2 there are N_2, and the parameter defining their susceptibility is a_2. And so on, with as many classes as are needed to describe the wounds. No limit is put to the number of classes. Then we must replace the Poisson equation by

$$Y = N_1(1 - e^{-a_1 x}) + N_2(1 - e^{-a_2 x}) + \ldots \qquad (2.2)$$

In this equation no allowance has been made for the nonrandom distribution of spores. If in a wound of class 1, there are $b_1 x$ spores, in a wound of class 2, $b_2 x$ spores, and so on, the effect is merely to increase the total number of classes; and the model allows for this because it does not call for a limitation on the number of classes. And if b_1, b_2, . . . are independent of x, their inclusion in Eq. (2.2) would not alter any deductions, and we exclude them in the interest of simplicity.

When $a_1 x$, $a_2 x$, $\cdots$ are all small the approximation used to derive Eq. (2.1) can be used to reduce Eq. (2.2) to

$$Y = (N_1 a_1 + N_2 a_2 + \cdots)x \qquad (2.3)$$

This too is an equation of a straight line passing through the origin, no different in this respect from the simpler Eq. (2.1), When the number of susceptible sites (e.g. wounds in the case of green mold of oranges) is very large compared with the number already infected, so that a_1x, a_2x, . . . are all small, the variety of the susceptible sites is irrelevant. Equation (2.3) is the equation of Curve A of Fig. 1.7; and Curve A will accommodate almost any degree of diversity and heterogeneity provided that the number of susceptible sites is very great.

We must examine more closely the proviso that the number of susceptible sites must be great. Suppose susceptible sites of class 1 are more easily infectible than those of class 2, and so on with classes arranged in decreasing order of infectibility. Then $a_1 > a_2 > $ . . ., and $a_1x > a_2x > $ . . . . In order to use the approximation that changes Eq. (2.2) to Eq. (2.3) it is therefore necessary only that a_1x should be small. One needs to consider only the most easily infectible sites. If they are many, with only a small proportion infected, Eq. (2.3) will hold, and the disease/inoculum curve will be (within the limits of the approximation) a straight line, as in Curve A of Fig. 1.7. But as soon as these most easily infectible sites become more heavily infected, so that the competition among spores for them is significantly great, the disease/inoculum line will curve to the right, as in Curve B.

Equation (2.3) has in reality only one parameter, $N_1a_1 + N_2a_2 + \cdots$, although this parameter is a composite parameter made up of an indefinite number of contributing parameters. This parameter defines the slope of the disease/inoculum line. (It is the tangent of the angle the line makes with the x-axis.) Thus, in the data on bean rust in Fig. 1.1, there was an average of 1 lesion per 11 spores, so the parameter was $1/11$. For the purpose of this particular chapter we need dig no deeper.

The Poisson equation—Eq. (1.2)—has two parameters, a and N. and its curve is defined by them. But its modification—Eq. (2.2)—has many, indeed countless, parameters; and the form of the disease/inoculum will vary in countless ways. The only generalization that seems possible now (except near the origin) is that the curve will have no maximum—i.e., no point of inflexion—as distinct from an asymptote. This distinguishes Curve B from Curve C of Fig. 1.7.

Consider some examples. Figure 1.1 relates the number of lesions to the number of spores of *Uromyces phaseoli*/cm^2 of bean leaf. The line is straight and passes through the origin. Its equation therefore has only a single parameter, which we identify with composite parameter in Eq.

(2.3). This equations holds only when $a_1, a_2, \ldots$ are small, and by implication $N_1, N_2, \ldots$ are large. The line is straight for the whole range of the data, with the number of lesions up to about 80/cm² of leaf. This represents a very high incidence of disease, greater than is likely ever to be found in nature. From this it is clear that the number of susceptible sites, $N_1, N_2, \ldots$ was large, at least of the order of thousands/cm². This is to be expected, because *U. phaseoli* can penetrate a bean leaf at almost countless susceptible sites. Also, the larger the number of susceptible sites, the smaller is *a*, because *a* reflects the probability that a spore will infect a given site. For simplicity of illustration consider the matter in terms of the Poisson equation (2.1) instead of the more realistic equation (2.3). From the slope of the line in Fig. 1.1 it appears that the probability that a spore will infect one or other susceptible site is about 1/11, whence *a* is $1/(11N)$ and therefore very small.

Figure 1.2 illustrates Curve *B* of Fig. 1.7. Curvature becomes apparent even when the number of lesions caused by *Botrytis fabae* is only about two per half-leaflet of *Vicia faba,* which contrasts with Fig. 1.1 in which the disease/inoculum line was still straight when there were 80 lesions per cm². Under the conditions of the two experiments on which Figs. 1.1 and 1.2 are based, *B. fabae* was much less successful in finding susceptible sites than *U. phaseoli.*

A more extreme form of Curve *B* of Fig. 1.7 is shown by the results of Leach and Davey (1938) for *Sclerotium rolfsii* on sugar beet. They related the percentage of sugar beets infected on August 1 with the number of sclerotia per square foot of soil to a depth of 8 in. The relation is shown in Fig. 2.1. Disease increased fast with the concentration of sclerotia until almost 20% of the plants were infected. Then the curve turned sharply and flattened out. Susceptible sites were severely limited, and restricted to about 25% of the plants. For these few sites competition was strong as judged by the marked curvature.

The next two examples illustrate the interplay of *N* and *a*. Low *N* values or high *a* values increase curvature, so a combination of low *N* and high *a* values should markedly increase curvature.

McKee and Boyd (1952) inoculated potato tubers with *Fusarium caeruleum,* a wound parasite. They made standard holes in tubers, and inoculated the holes with spores suspended in a drop of water. In interpreting their results we must at the start accept that a hole in the tuber is not in itself a susceptible site in the meaning of this section. Many holes remained healthy despite massive doses of inoculum. Thus when

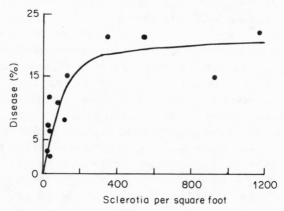

Fig. 2.1 The relation between the percentage of diseased sugar beets and the number of sclerotia of *Sclerotium rolfsii* per ft² of soil to a depth of 8 in. (Data of Leach and Davey, 1938.)

McKee and Boyd used sprouting tubers, infection reached a plateau or asymptote far short of 100%, resembling except in degree the results with *S. rolfsii* on sugar beet. With 30, 60, 121, and 242 spores per infection drop the percentage of holes infected remained constant at 74, 75, 74, and 73, respectively; one-fourth of the holes were evidently highly resistant to infection. In Fig. 2.2 we convert the percentage of holes infected into the estimated average number of infections per hole, this estimated average being m in Eq. (1.1). This allows, as best we can direct, for multiple infections of a hole. Figure 2.2 shows a strong contrast between dormant and sprouting tubers. The disease/inoculum curve for sprouting tubers levels out at about 1.3 infections per hole (which corresponds with 75% of the holes infected). Clearly, the number of susceptible sites was limited, i.e., N values were low. With dormant tubers on the other hand disease increased continuously with increasing inoculum, and continued to do so for inoculum doses outside the range of the graph. N values in dormant tubers were relatively very high. When attention is switched from the upper parts of the curves to parts near the origin, the disease/inoculum curve for sprouting tubers is found to be steeper than for dormant tubers. (This part of the graph, near the origin, is treated in more detail in our next example, and is not elaborated on here.) Thus $N_1a_1 + N_2a_2 + \ldots$ for sprouting tubers is greater than for dormant tubers; and because the N values for sprouting tubers were lower we can reasonably deduce that the a values were considerably higher. In plain language, the number of susceptible sites

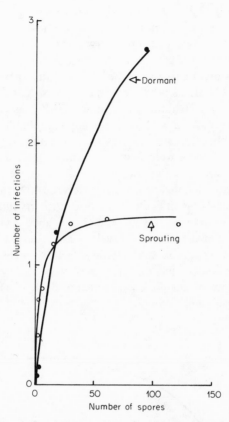

Fig. 2.2   The relation between the estimated number of infections of potato tubers and the number of spores of *Fusarium caeruleum*. The number of infections was estimated as *m* in Eq. (1.1). (Data of McKee and Boyd, 1952.)

in holes in sprouting tubers was lower but individually they were much more easily infectible. This combination made for the greatly increased curvature of the disease/inoculum curve for sprouting tubers.

Figure 1.18 reproduced some data of Ercolani (1967) on the artificial infection of tomato plants by *Corynebacterium michiganense*. We must now take another look at that figure. With the tomato varieties Roma and Fiorentina the disease/inoculum curve is practically straight. It is of the Curve *A* type. There were many susceptible sites; the *N* value were high. But the variety C1402 gave an entirely different response. The disease/inoculum curve is markedly bent, and is of the Curve *B* type. In quickly approaching a plateau it resembles *S. rolfsii* on sugar beet in Fig. 2.1 or *F. caeruleum* on sprouting potato tubers in Fig. 2.2. The number of susceptible sites is very low; the N values are very low. In Fig. 2.3 we repeat on a larger scale, details near the origin. With the variety of C1402 curvature begins very soon. The slope of the disease/inoculum curve near the origin seems to be about 15–20 times as great for C1402 as for Fiorentina. That is, $N_1a_1 + N_2a_2 + \ldots$ seems to be about 15–20 times as great for C1402 as for Fiorentina. With N values lower for C1402 it seems not improbable that a values were hundreds of times greater. In C1402, under the conditions of Ercolani's experiments, there were fewer susceptible sites, but individually they were more easily infectible, hence, the sharpness of the curvature. Until one knows how homogeneous the variety C1402 was—was it genetically a mixture?—it is difficult to analyze further. But, whatever its constitution may have been, C1402 remains a good example of how high a values combined with low N values cause strong curvature.

The two-path concept of susceptibility clearly needs much more probing before we understand the principles of plant infection. Meanwhile, in the absence of adequate experimental data, it remains only to assess some practical implications.

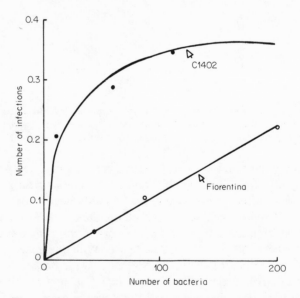

Fig. 2.3 The relation between the number of infections of tomato plants and the number of cells of *Corynebacterium michiganense* in the inoculum. This figure magnifies part of Fig. 1.18. (Data of Ercolani, 1967.)

2.3   THE RELEVANCE OF THE TWO-PATH CONCEPT OF SUSCEPTIBILITY TO MODELS IN EPIDEMIOLOGY

Curve *A* of Fig. 1.7 is the disease/inoculum curve that has been assumed in all mathematical models of epidemics discussed in the literature. Its advantage is great. Disease response is directly proportional to inoculum dose. To answer the question at the beginning of Chapters 1 and 2, if 1000 spores fall on plants and cause *n* lesions, then with Curve *A*, 2000 spores would have caused an average of 2*n* lesions. That sort of answer fits easily into a model. But with Curve *B* the answer is somewhere between *n* and 2*n* lesions, the number varying not only with the effect of the whole disease triangle of host, pathogen, and environment on all the parameters of Eq. (2.2) but also with the percentage of disease that *n* lesions represent. That sort of answer makes quantitative models unworkable in our present state of knowledge, except at very low levels of disease. (With Curve *C*, the answer is less than 2*n*; with Curve *D* it is more than 2*n*. Both answers make quantitative models unworkable at present.)

Curve *A* means that all the parameters in Eq. (2.2) are dissolved into a single composite parameter. In using Curve *A* we use a mathematical relation that probably does more to simplify epidemiology than any other. In terms of the two-path concept of susceptibility the one condition for this simplification is that the number of susceptible sites should be very large in relation to the number that become infected.

Is it justified generally to assume that Curve *A* commonly, but not invariably, holds for obligately parasitic fungi? Possibly obligate parasitism, implying as it does great adaptation of pathogen to host, means that the fungus has little difficulty in finding or making susceptible sites. Figure 1.1 concerned *Uromyces phaseoli*. Figure 1.4 concerned *Puccinia graminis tritici*, and the relevant points (as distinct from the later points which we shall suggest are irrelevant artifacts) are in a straight line. This result with *P. graminis tritici* was obtained also by Rowell and Olien (1957). Some artifacts excluded, the data on *Erysiphe graminis* to be discussed in the next section fit Curve *A*. Figure 1.9 concerned *Synchytrium endobioticum*. The lines in Figs. 1.13 and 1.14, for *Hemileia vastatrix* and *Tranzschelia discolor*, respectively, do not depart significantly from straightness, but the data are too few to carry much weight. On the other hand, it is possible to quote examples that depart greatly from Curve *A*. Colhoun's (1961) data on the incidence of club root of cabbages caused by *Plasmodiophora brassicae* give a sharply bent curve, with disease and inoculum far from being directly proportional. Here we are dealing with a disease very sensitive to variable environmental conditions and host nutrition, and these may dominate host–parasite relations in the disease triangle.

Curve A often holds for the data on the more specialized nonobligate parasites. It holds for Knutson and Eide's (1961) data for *Phytophthora infestans* on potato plants, up to concentrations that produce 250 lesions per plant. The potato plants were small (only 30–38 cm high); and 250 lesions per plant represents practically complete infection. It holds for McCallan and Wellman's (1943) data for *P. infestans, Alternaria solani,* and *Septoria lycopersici* on tomato leaves. It holds for Segall and Newhall's (1960) data about *Botrytis allii* on onion. Among bacteria it holds for the examples discussed in Section 1.13.

There are many exceptions, and even with the same disease results vary, Curve A being found in some experiments and Curve B in others. In one experiment of Wastie (1962) with *Botrytis fabae* on *Vicia faba* the disease/inoculum curve did not depart significantly from straightness, i.e., it did not differ significantly from Curve A, after the data were corrected for multiple infection. But in a second experiment the departure was marked. The curve was Curve B; and, after corrections for multiple infection, the curve did not differ much from that applicable to Last and Hamley's (1956) results with the same disease. It so happens that Curve A in the first experiment was associated with a lower ED_{50}, but it is doubtful whether there is much significance in this. In the second experiment with *B. fabae*, giving Curve B, the curvature (as distinct from the slope) of the disease/inoculum curve was much the same when *B. cinerea* replaced *B. fabae*, but the ED_{50} with *B. cinerea* was much higher. A lower ED_{50} would suggest a straighter disease/inoculum curve if the greater susceptibility of the host (or greater aggressiveness of the pathogen) was due to more numerous susceptible sites, but not if it was due to the susceptible sites being individually more easily infectible.

The specter of the possible irrelevance of artifacts haunts much of the literature of disease/inoculum relations; and one can endorse without reservation Carter's (1972) plea for more realism. Carter deplored the lack of information about the configuration of spore deposits of fungus pathogens on host tissues, and on the redistribution of spores by rainfall or overhead irrigation; and he calls for a technology for studying the ultimate fate of deposited spores. Carter reminds us that spores of most "dry-dispersed" pathogens, e.g., rust fungi, powdery mildew fungi, *Monilinia* spp., and *Botrytis* spp., are hydrophobic and tend to be liberated in clumps of various sizes. The slimy spores of pathogens such as *Fusarium* spp., *Gloeosporium* spp., or *Ascochyta* spp. are hydrophilic and normally splash dispersed during rainfall or irrigation. They too are distributed in groups. Whether hydrophobic or hydrophilic, inoculum seldom arrives at the host surface as uniformly spaced spores. Except for those of the powdery mildews, spores of most fungus pathogens are believed to need contact with a water droplet before they can germinate. But this does not mean they need to be buried in water. In artificial inoculations for the study of disease/inoculum relations it is common practice to use a uniform, standardized suspension of spores in an aqueous medium; and the experimenter is seldom satisfied until he has succeeded in breaking up the spore clumps, thoroughly wetting hydrophobic spores that normally might never be wetted and burying them in water. The effect of this was studied by Carter. He used uredospores of the rust fungus *Tranzschelia discolor*. Of those suspended in 0.06% agar and kept buried deep, 0.5% germinated. Of those delivered to water–agar in a Petri dish and allowed to rest on the surface, 16% germinated.

2.4 CURVE *C* OF FIG. 1.7. THE ANTAGONISTIC INTERACTION OF SPORES

This topic was introduced in Section 1.4. Davison and Vaughan (1964) added a suspension of uredospores of *Uromyces phaseoli* to bean leaves, held the plants in a moist chamber for 16 to 18 hours and then in a greenhouse for 2 weeks, after which they counted the pustules. Figure 1.3 relates the number pustules/cm^2 of leaf surface to the number of spores/cm^2. The first two points are in line with the origin, and the pustules they represent were normal in appearance. The third point, with about 27 pustules/cm^2, reflects unnaturally high concentrations of disease. The pustules were crowded and below average size, and unlike the pustules that develop in natural epidemics. With still greater numbers of spores the number of pustules decreased sharply, considerably fewer pustules being formed by 3400 than by 1100 spores/cm^2.

Because of the overcrowding of pustules, with more than 25/cm^2, and their abnormal appearance, we regard the results with high concentrations of spores as artificial. Added to the artificiality of abnormally high numbers of pustules was the artificiality of spores settling on the leaves in a single, simultaneous salvo; in natural epidemics spores settle over days or weeks, so that at any one time the concentration of fresh spores is relatively low. In relation to epidemiology the last three points on the graph lack realism; they are irrelevant artifacts.

The same sort of result was obtained by Domsch (1953) for *Erysiphe graminis* on barley. He suspended spores in water and applied them to the leaf surface. Disease was directly proportional to inoculum, with 1.6 pustules/1000 spores, until the inoculum contained 2×10^5 spores/ml. That is, up to this stage the disease/inoculum curve was Curve *A*. Thereafter, the proportion began to decrease; and the amount of disease reached a maximum of about 1.5 pustules/cm^2 of leaf and then declined. Here again the decline began only after unnaturally high levels of disease had been reached, after inoculation with a simultanous salvo of spores.

Curve *C* is apparent also in Khan's (1972) experiments with infection of cotton roots by *Fusarium vasinfectum,* again only at very high doses of inoculum. The amount of inoculum was measured in arbitrary units. With the cotton variety Karunganni–5 disease started to decrease when the amount of inoculum was increased to five times the dose needed to cause 100% infection. With the variety 320–*F* disease also started to decrease when the amount of inoculum was five times the dose needed to cause maximum infection, which was 80–82%. With three other pathogens, *Rhizoctonia solani, Sclerotium rolfsii,* and *S. bataticola,* the curve was Curve *B*.

2.5   Curve *D* of Fig. 1.7. The Synergistic Interaction between Spores

The topic was introduced in Section 1.5. Petersen (1959) allowed dispersed uredospores of *Puccinia graminis tritici* to fall on wheat plants and infect them. After an appropriate interval in an infection chamber and greenhouse he stained the leaves and counted the infection points. The counts are recorded in Fig. 1.4, to which we must now refer in more detail. There are 15 points in the figure. They are divided into two groups. The first 9 points up to 2810 spores and 330 infection points/cm^2 have been treated as a group, separate from the last 6 points, up to 5640 spores and 1520 infection points/cm^2. The first 9 points are more or less in a straight line—departures from straightness are statistically insignificant. At this stage there is no evidence of synergism, i.e., there is no tendency for the disease/inoculum curve to turn upward. The last 6 points show a sharp upward turn. The regression coefficient of the line fitted to them is 0.490 which compares with a regression coefficient of 0.108 for the line fitted to the first 9 points; and the difference is significant statistically. There were about 620 infection points with 3600 spores/cm^2, a ratio of 17:100, and 1470 infection points with 5400 spores/cm^2, a ratio of 27:100. At high concentrations the spores helped one another to infect; they acted synergistically. There is other evidence for this. High concentrations of spores stimulated germination. With 850 spores/cm^2, 28% of the spores germinated and produced appressoria; with 5300 spores/cm^2, 52% germinated and produced appressoria.

The division of Fig. 1.4 into two parts has another purpose. The first 9 and the last 6 points separate themselves out on the modified Cobb scale commonly used to assess wheat rust. The first 9 points, up to 2810 spores/cm^2, are compatible with field epidemics. These 2810 spores/cm^2 produced 330 infection points which in turn produced 18 pustules/cm^2; and 18 pustules/cm^2 represent 100% infection on the Cobb scale. The last 6 points are incompatible with field epidemics. They represent superinfection, above 100%. This artifact was made possible because the spores arrived in a single simultaneous salvo and the infection points were stained and counted before, through sheer overcrowding, they fused together into larger and fewer pustules.

Here is incontrovertible evidence that the synergism represented by the last 6 points was an artifact. We are left with the first 9 points. There is evidence for Curve *A* of Fig. 1.7; but, excluding artifacts, none for Curve *D*.

We must scrutinize this evidence carefully. Petersen's technique separated the uredospores as much as possible (he assures us that the spores settled as individual units); and his evidence with low concentrations of inoculum was against synergism between spores which were the units of dispersal. But had the spores been allowed to settle in large clumps, i.e., had clumps of spores been the units of dispersal, Petersen's evidence for spores at high concentrations suggests that there would have been synergistic action between spores within clumps. Synergism within units of dispersal, as distinct from between units of dispersal, does not give a disease/inoculum curve of the Curve *D* type. It gives Curve *A* or Curve *B*, according to the parameters. What it does do is to alter the slope of the curve, as shown by the angle the disease/inoculum line makes with the x-axis. The slopes of curves are a matter for Chapter 3.

Curve D implies facultative synergism. But the converse is not necessarily true. Facultative synergism does not always imply Curve D. Consider the results of Kelman and Sequeira (1965) with *Pseudomonas solanacearum*. Massive populations of bacteria, with 3.2×10^8 cells/ml of soil water, were able to infect healthy unwounded roots. Kelman and Sequeira suggest that these massive doses can enzymatically digest the mucilagenous sheath or other barriers to infection of the root, and infection can spread from plant to plant. This is evidence for synergism, in that massive populations of cells can infect in a particular way that dilute concentrations of cells cannot. But dilute concentrations have their own way of infection (as witness, e.g., the difficulty farmers have of getting rid of infection in the land). To prove that Curve D pertains, it would be necessary to prove that 3.2×10^8 cells/ml cause more than 3.2×10^8 times as many infections as 1 cell/ml would do. That is, it would be necessary to prove that the advantage of concentration outweights the advantage of dispersal. When spores of *Puccinia graminis* occur at high concentration they become individually more effective as pathogens: A higher percentage germinates and forms appressoria. This is the reason for Curve D (at least in artificial conditions). But there is no corresponding evidence that high concentrations of *Pseudomonas solanacearum* make each cell individually produce more of the enzymes that dissolve host mucilage or other barriers to infection. Until evidence is available we must keep an open mind about Curve D here.

2.6   A Short Summary of the Chapter So Far

To summarize briefly, in Chapter 1 four experimental disease/inoculum curves were discussed. Curve A of Fig. 1.7, a straight line through the origin, indicates that spores (or units of dispersal generally) act independently of one another; there is neither competition between spores for susceptible sites nor interaction between them. Curve B indicates competition for susceptible sites, but not interaction, antagonistic or synergistic, between units of dispersal. Curves C and D indicate antagonistic and synergistic interaction, respectively, between units of dispersal which in the experiments discussed were mainly or wholly individual spores. In Chapter 2 we regard Curves C and D as usually being artifacts though there may be occasional exceptions. It is not disputed that antagonism and synergism occur; the chemical and general evidence that they do occur is overwhelmingly strong. All that is implied in abandoning Curves C and D is that if antagonism and synergism do occur they are more likely to occur within clumps or other associations of spores, when these clumps or associations are the units of dispersal, than between the units of dispersal. Propinquity between spores within a unit of dispersal is enough to determine this. Interaction within units of dispersal does not produce Curves C and D; it produces Curves A and B.

Curves A and B are the fundamental curves. On the two-path concept of susceptibility, they can be derived from the same model simply by a change in the numerical values of parameters. Curve A implies a great number of susceptible sites in relation to the number of spores that could use them. Curve B implies a restricted number of susceptible sites and hence competition for them. In terms of constructing models or devising computer programs, Curve A has the immense advantage of having the various parameters merged into a single composite parameter. Curve B occurs commonly; and if in quantitative epidemiology it is urged that attention for the present be concentrated on diseases with Curve A, this is no attempt to escape from reality but simply advice to first learn how to walk before trying to run.

2.7   DISEASE/INOCULUM CURVES IN ROOT DISEASE

It has been claimed in the literature (Baker *et al.,* 1967; Martinson, 1963; Semeniuk, 1965) that with respect to disease/inoculum curves, root disease differs fundamentally from foliage disease. Soil (the argument runs) is a three-dimensional medium, and spores or other inocula in soil are distributed in three dimensions. A root surface is two-dimensional, and spores on the surface are distributed in two dimensions. Baker *et al.* (1967) have proposed mathematical models which they believe establish a difference. They believe in a logarithmic relation between disease and inoculum. Thus, with nonmotile spores and a moving infection court (i.e., growing roots) or with nonmotile spores about a fixed infection court they suggest that the logarithm of the amount of disease plotted against the logarithm of the amount of inoculum should have a slope of $2/3$.

We reject all this, for two reasons.

First, as a matter of experimental observation disease/inoculum curves in root disease do not differ from those in foliage disease. There is no known evidence that disease/inoculum curves for roots are in any way unique. Thus, Last and Hamley (1956) inoculated leaves of *Vicia faba* with spores of *Botrytis fabae*. (A few of their data are given in Fig. 1.2.) They showed from a massive amount of data that the logarithm of the number of lesions was proportional to the logarithm of the number of spores used as inoculum; and a further analysis of the log-log line shows that its slope does not differ significantly from ⅔. This is the slope that Baker *et al.* claim to be a characteristic property of root disease as a result of the three-dimensional distribution of spores in soil. Similarly in Wastie's (1962) experiments with spores of *B. cinerea* applied to the surface of leaves of *Vicia faba* the log-log slope was approximately ⅔. By selecting appropriate ranges of inoculum a log-log slope of approximately ⅔ can be obtained by rubbing suspensions of tobacco mosaic virus on leaves of *Nicotiana glutinosa*. Baker *et al.*'s models can also be condemned from another angle: They do not predict disease/inoculum curves for root disease that are in fact known. They do not allow for the number of infections to be directly proportional to the number of spores, as in Glynne's (1925) data for *Synchytrium endobioticum* on potatoes. See Fig. 1.9. They do not allow for the disease/inoculum relation found when *Sclerotium rolfsii* attacks sugar beet, illustrated in Fig. 2.1. But this relation found when *S. rolfsii* attacks sugar beet, far from being a relation characteristic of soil disease, is matched by the relation found when *Corynebacterium michiganense* attacks foliage of the variety C1402 of tomato. See Figs. 1.18 and 2.3.

Second, the models of Baker *et al.* are based on three assumptions that must be challenged. *One,* they assume spores to be mathematical points in the soil, without dimensions and volume. In reality spores have dimensions and volume; and this is entirely relevant, because the essence of the problem is, as we shall see, the volume of soil they occupy near the root. *Two,* they assume spores to be uniformly distributed in soil in such a way as to be situated at the vertices of regular tetrahedra. In reality, even after the most thorough mixing, spores would not be more than randomly distributed in soil; and random distribution is not in regular tetrahedra. *Three,* they assume the number of infections to be directly proportional to the number of spores at the root surface. In effect, they assume that infection is never restricted by the number of susceptible sites, and ignore the possibility that many spores would have to compete with one another for sites. They neither present evidence for their assumption nor discuss it.

To inject some realism into the topic we suggest a simple model needing no involved mathematics and dealing with nothing but volumes. Suppose there is a supply of infected soil, well mixed so that the (non-motile) spores are randomly distributed through it. This is the standard soil. Suppose further some of this soil is added to an equal volume of uninfected but otherwise similar soil, and well mixed so that the spores are again randomly distributed. The concentration of spores in this batch of soil will then be half that of the standard soil. And suppose further that a third batch of soil is made by mixing one part of the standard soil with two parts of uninfected soil. The concentration of spores in this batch will be a third of that of the standard soil. We now put each of the three batches of soil—standard strength, half-strength, and third-strength as regards concentration of spores—in pots. The average number of spores in a pot with standard strength soil is, say, n, so that the average number in a pot with half-strength soil is $n/2$, and the average number in a pot with third-strength soil is $n/3$. In each pot a seed is sown, and the seedling and its roots allowed to grow until 1% ·of the volume of the soil is within the rhizosphere. The volume of the spores being an integral part of the volume of the soil, an average of $n/100$ spores will be within the rhizosphere of a plant growing in a pot of standard soil, $n/200$ within the rhizosphere of a plant growing in a pot of half-strength soil, and $n/300$ within the rhizosphere of a plant growing in a pot of third-strength soil. The number of spores within the rhizosphere is in simple proportion to the number in the pot. No logarithms are needed to describe this relation. The further relation between the number of spores within the rhizosphere and the number of infections they cause could follow either Curve A of Fig. 1.7 or Curve B in any of its various forms. If there were very many susceptible sites on the roots, one would expect Curve A; if there were few susceptible sites with many spores having to compete for them one would expect Curve B.

The rhizosphere, like the rest of the soil, is three-dimensional; it has volume. As soon as one discards Baker *et al.*'s abstraction of spores as mathematical points without volume, and recognizes spores as occupying real volume of soil, the argument about three-dimensional soils and two-dimensional surfaces falls away.

In the model one can replace the rhizosphere by any other volume of soil in which a spore must lie if it is to be in a position to attack a root; and one can change to motile spores. Thus, if the spores are motile and attracted by root exudates to the root, one must allow for a thicker hollow cylinder about the roots from which the attack could come. We need not know the volume of soil in the rhizosphere or hollow cylinder; without change of argument the 1% in the model could be changed to an unknown X%. All the model requires is that the amount of root in each pot should be the same. And because the model holds for any chosen value of X it holds for growing roots, i.e., for a moving infection court.

We retain Curves A and B of Fig. 1.7 for both root disease and shoot disease, despite the differences between soil and air as environments. Soil is mostly dark and sunless. It is better buffered against changes of temperature and moisture. It contains relatively great amounts of the dead remains of plants and animals, and of excretions of living plants and animals; these serve both to maintain a large microbial population living saprophytically and to act as food bases from which parasitic attack can be launched. Competition and interaction between pathogens and other microorganisms in soil is more intense than on leaves. Many spores in soil—sexually produced oospores and asexually produced clamydospores or multicellular sclerotia—are adapted to dormant survival. In addition to an endogenous dormancy there is an exogenous dormancy—fungistasis—imposed by the environment of the soil. In shoot disease the pathogen comes to the host; and the aerial movement of spores, with or without water, is well developed. In root disease the pathogen may also come to the host; aerial movement is more restricted, but mycelial spread, e.g., of *Pythium* spp. or *Rhizoctonia solani* has the mechanical support of soil. In addition, in root disease the host goes to the pathogen, when roots spread in infected soil and reach pathogens such as *Fusarium* spp. that simply lie in wait. In root disease ectotrophic infection is more common than in shoot disease (though it has possibly been underrated in shoot disease); the fungus spreads superficially over the root surface before infecting. In root disease, fungus aggregations in the form of infection cushions, mycelial sheets, strands, and rhizomorphs are frequent. Root exudates from host plants are relatively more important, usually, than leaf exudates; they attract growing pataogens or swimming zoospores towards the root and nourish them. In root disease, toxic exudates from pathogens are relatively more important; they weaken or kill host tissue in advance of infection. In root disease, environmental pollution—particularly the production of phytotoxins by decomposing plant residues—was relatively more important than in shoot disease, though the modern surge of atmospheric pollution may well have now reversed this.

These differences do not alter the fact that in root disease as in shoot disease infection depends on the quantity and quality of susceptible sites. If sites are many, the disease/inoculum curve will be Curve A of Fig. 1.7. If sites are few and spores or other pathogenic material must compete for them, the curve will be Curve B, and details will be determined by the interplay of the quantity and the quality of the sites.

2.8 TESTS FOR INDEPENDENT ACTION OF SPORES

We have discussed simple tests. Curve A of Fig. 1.7 is for the independent action of spores. Curve B is for competition between spores for susceptible sites. Curve C is for antagonistic interaction between spores, and Curve D for synergistic interaction. These curves are on simple arithmetic axes. But the literature discusses more complicated tests to which writers commonly refer. A short survey is needed.

Druett (1952) and Peto (1953) showed that when homogeneous test animals are exposed to infection by microorganisms there should be a linear relation between the logarithm of the proportion of survivors, i.e., animals not infected, and the number of microorganisms, if the microorganisms infect independently. The proportion of survivors is $1 - y$, which has the same meaning as in Eq. (1.1). Druett and Peto plot $+\log_e (1 - y)$ against the number of microorganisms, and the ordinates are necessarily negative. In the examples considered in Chapters 1 and 2, I plotted either the observed number of infections where this was recorded in the literature or the estimated number where disease was recorded as the percentage or proportion of plants infected. This estimated number m is calculated by Eq. (1.1) as $-\log_e (1 - y)$. Druett and Peto's test and my test are in effect the same. They plot $+\log_e (1 - y)$ with ordinates negative; I plot $-\log_e (1 - y)$ with ordinates positive. The two tests are identical except that the signs are interchanged. My test has two advantages. It can be used, whereas Druett and Peto's test cannot be used, when disease is expressed as a counted number of lesions (per plant, per leaf, per cm^2, or on any other basis). It has biological simplicity, in that it relates the observed or estimated number of infections directly to the number of spores (or microorganisms) that cause them; it sees the problem simply in terms of disease response to inoculum dose.

Peto (1953) devised another test. The probit of the proportion of infected plants—which we shall call probit disease—on the y-axis is plotted against the logarithm of the number of spores—which we shall call log dose—on the x-axis. If the dose is expressed by multiples of ED_{50}, then the slope of the curve at ED_{50} will be approximately 2, if spores do not interact. (The logarithms here are to the base 10.)

This complicated test for interaction has come to be regarded by plant pathologists as the standard test, and has been widely used. But the extra complication brings no extra reward; the test does nothing that could not be done more simply otherwise. Peto brought in this complicated test—that the slope of the probit-disease/log-dose curve at ED_{50} should be approximately 2— as an incidental in the course of demonstrating the value of the simpler log-survival/dose relation. It testifies to a widespread love of display that the simpler and better test lies buried forgotten in the literature, whereas the complicated test, mathematically clumsy, biologically inept, and understood by few, appears regularly in publications.

The tests for independence, competition, and interaction illustrated by Curves $A, B, C,$ and D of Fig. 1.7 combine mathematical simplicity with biological clarity. They are both more versatile and more comprehensive than either the log-survival/dose or probit-disease/log-dose methods; and there appears to be no good reason why the older, more complicated tests should not disappear from future literature.

2.9   THE USE OF LOGARITHMS IN DISEASE/INOCULUM STUDIES

Parris (1970) states that it is an accepted principle in the literature that the amount of disease varies with the logarithm of the amount of inoculum. Parris is possibly right in stating that this is a widely held belief among plant pathologists; but the belief itself is totally wrong.

Nowhere in this book, in any graph, is the logarithm of the amount of inoculum used. The first graph in the book, Fig. 1.1, sets the pattern. On arithmetic axes the number of lesions/cm^2 of bean leaf on the y-axis is plotted against the number of spores/cm^2 on the x-axis. The relation is crystal clear: Disease is proportional to inoculum, and there is one lesion for approximately every 11 spores.

There are several objections to using logarithmic axes in disease/inoculum studies.

First, logarithms obscure relations. Readers can test this for themselves. Take Fig. 1.1, just discussed, and change to the logarithm of the number of spores on the x-axis. The whole relation is now obscure; and most of us would first have to delogarize the graph, mentally or with tables, to get the message that disease was proportional to inoculum and there was one lesion for every 11 spores.

Second, because logarithms obscure relations they are likely to be misread. Last and Hamley (1956) applied spores of *Botrytis fabae* to leaves of *Vicia faba* and counted the lesions they formed. They found in repeated experiments that the logarithm of the number of lesions was directly proportional to the logarithm of the number of spores, and they concluded that the number of lesions was directly proportional to the number of spores. This conclusion was rated important enough to be included in the short abstract that appeared at the head of their paper in the *Annals of Applied Biology;* it was accepted by the two authors, the journal's editor, and presumably by an unknown number of referees; and it has been taken up in a standard work on plant pathology. But the conclusion makes no sense mathematically. It does not follow, that because the logarithm of the amount of disease is proportional to the logarithm of the amount of inoculum, disease is proportional to inoculum. A table of antilogarithms applied to the data of Last and Hamley shows disease/inoculum curves bending strongly to the right; that is, inoculum becomes proportionately less effective in producing disease as the amount of inoculum increases. (Some of their data were used in Fig. 1.2.) The curves are examples of Curve *B* of Fig. 1.7.

Let it be explained that only in one special circumstance is disease directly proportional to inoculum when the logarithm of disease is directly proportional to the logarithm of inoculum, and that is when the slope of the log–log curve is 1 (the scales on both axes being the same). In Last and Hamley's data the slope was roughly $\frac{2}{3}$. When the slope of the log–log curve is less than 1, the disease/inoculum curve bends to the right; and when the slope is as low as $\frac{2}{3}$, it bends strongly to the right.

A failure to read logarithms correctly that has misled workers badly is discussed in Section 4.7.

Third, with arithmetic scales the graph has a hitching point: The curve is hitched at the origin (see Section 1.11). Curves of logarithmic data float unhitched. Biological data are notoriously uncertain, and it is folly to discard the one point of reasonable certainty.

Logarithms have an advantage. They can be used to compress data that stretch over a wide range, as when, e.g., the number of spores used as inoculum varies in an experiment from one to a million. But one pays for that advantage. As data are compressed, details are correspondingly suppressed. The processes are concomitant. Massive ranges of data compressed by logarithms into small compass are unquestionably often useful in revealing gross trends; but if one wants to examine detail it is useful to break the data up into less massive ranges, and then to examine them delogarized.

This section is concerned with logarithms in disease/inoculum relations and, especially, with logarithms of the amount of inoculum. Here logarithms are useful only for compression and indirectly in estimations. But the comments in this section in no way detract from the usefulness of logarithms where they are appropriate. The criticism of the unnecessary use of logarithms is intended for the particular field of study that falls within Chapters 1 and 2, and not generally.

2.10 PARAMETERS IN DISEASE/INOCULUM RELATIONS

When disease is directly proportional to inoculum the disease/inoculum line is straight and passes through the origin, as in Fig. 1.1. Only one parameter is needed to describe the relation. But when there is curvature two or more parameters are needed, and it is appropriate to review some problems that arise. First, there is the problem of increasing the number of parameters. Second, there is the problem of finding appropriate parameters. And, third, there is the difficulty of finding the error of any particular parameter when there are many of them.

For illustration the data in Table 2.1 are used. These are Kleczkowski's (1950) data for the number of local lesions formed when tobacco mosaic virus solutions of varying concentration are rubbed on to leaves of *Nicotiana glutinosa*. (The data are for his Experiment 13.) What disease/inoculum equation, i.e., what equation relating the number of lesions to the concentration of virus, fits the facts and how does one interpret it?

Kleczkowski first fitted the Poisson one-hit equation—Eq. (1.2), repeated earlier in this chapter—to the data, using statistical methods to get the best estimates of the parameters, judged by the best fit they gave. There are two parameters in the Poisson equation, N and a, where N is the number of susceptible sites (per half-leaf), the susceptibility of all leaves being uniform, and a characterizes the susceptibility. The equation, Kleczkowski found, fitted the data badly; it failed to describe the disease/inoculum relations adequately.

Kleczkowski fitted another equation which, unlike the Poisson equa-

TABLE 2.1

Numbers of Lesions Obtained by Inoculating Tobacco Mosaic Virus at Different Concentrations [a]

Virus concentration (g/liter)	Lesion numbers on 12 half-leaves
0.000256	146
0.00128	276
0.0064	628
0.032	972
0.16	1484
0.8	1626
4.0	2510
20.0	2906

[a] Data of Kleczkowski (1950).

tion, allows for variations in susceptibility between the susceptible sites. The equation is

$$Y = \frac{N}{\lambda\sqrt{2\pi}} \int_{-\infty}^{t} \exp\left[-\frac{1}{2}\left(\frac{t-\xi}{\lambda}\right)^2\right] dt,$$

This equation assumes that the susceptible sites vary in susceptibility in such a way that the logarithms of the minimum virus concentrations necessary to cause a lesion to form are normally distributed. In the equation Y is the expected number of lesions per half-leaf, N is the mean number of susceptible sites per half-leaf, $t = \log_{10}x$, where x is the virus concentration of the inoculum, $\xi = \log_{10}x_0$, where x_0 is the virus concentration when half the susceptible sites develop lesions, and λ is the standard deviation. This equation has three parameters, N, ξ, and λ. It fitted the experimental data better than the Poisson equation did.

How does one interpret this better fit? Other things being equal, a three-parameter equation will give a better fit than a two-parameter equation, because there is an extra parameter, with which to make adjustments. The fact that the two-parameter Poisson equation did not adequately fit the experimental data leaves no doubt that the equation inadequately described the data both mathematically and biologically. But the better fit of the three-parameter equation leaves open the question whether the biological content of the equation was any better than that of the Poisson equation. Did the extra parameter add any biological reality, or was it simply one extra parameter with which to juggle empirically? In the actual process of fitting an equation to experimental data the parameters become mere ciphers to be adjusted so as to get the best fit. Whether the parameters reflect any biological truth is a matter that recedes into the background. It is most hazardous to attempt to extract biological significance from the finding that a three-parameter equation fits the data better than a two-parameter equation.

To leave biological implications aside for the moment, an extra parameter must be appropriate if it is to be useful, even when judged purely empirically. The three parameters in Kleczkowski's equation are highly appropriate. One of the parameters ξ ties the equation to the data at a point near which the data are abundant. The Poisson equation lacks this useful tie.

The appropriateness of parameters can only be judged in relation to what data are being analyzed. Kleczkowski's data were for wholly artificial conditions, in the sense that they did not describe any natural epidemic or situation that would arise in nature. He applied a solution of tobacco mosiac virus to an incompatible (i.e., naturally resistant) host, and with the aid of an abrasive infected it. The number of lesions increased with increasing virus concentration until an asymptote was reached; the highest number of lesions he produced was about the highest possible. At the other end of his data in Table 2.1 the lowest count was 146 lesions on 12 half-leaves, or about 12 per half-leaf. In a natural epidemic of tobacco mosaic virus on tobacco or other compatible host, even an average of one infection per whole plant would represent a moderately high level of disease in the field. To such natural conditions, with perhaps one susceptible site in a million infected from outside, Kleczkowski's equation would be entirely inappropirate. It would not even be possible to estimate parameters needed for the equation. Thus an equation which, empirically or otherwise, was appropriate to the data of a set of artifacts would be inappropriate to data collected during a natural epidemic. This is no criticism of Kleczkowski's equation in relation to the use to which he put it. He was concerned with the assay of virus concentration, and not with epidemics of tobacco mosaic in the field. Appropriateness must be judged by that. But by the same token Kleczkowski's criticism of the Poisson equation is valid only for his particular set of artifacts, and does not bear in any way on the relevance of this equation to natural epidemics of virus disease in the field.

Increasing the number of parameters in an equation increases the probability of getting a good fit to experimental data, provided that the parameters are appropriate and even if they are only empirically appropriate. What is the danger in increasing the number of parameters? Kleczkowski also used a four-parameter equation.

$$Y = N_1(1 - e^{-a_1 x}) + N_2(1 - e^{-a_2 x})$$

This is the same as Eq. (2.2), in Section 2.2, cut down to two terms. The four parameters are N_1, N_2, a_1, and a_2. They fitted the data quite well. Kleczkowski's estimates of the parameters for the data in Table 2.1 were $N_1 = 59$, $a_1 = 335$, $N_2 = 153$, and $a_2 = 0.63$. Note that a_1 is much greater than a_2; 532 times as great. It is possible that this difference is real, and that there are two very different processes of susceptibility involved in the infection of *Nicotiana glutinosa* by tobacco mosaic virus. But there is no biological evidence to confirm this, despite the fact that the infection of *N. glutinosa* has been well studied by many workers. One may therefore legitimately remain skeptical about the biological significance of the estimates. They may not be estimates of parameters appropriate to real susceptible sites, but simply estimates that balance one another to give the best possible fit.

With more than one parameter, there is danger of counterbalancing error, and the danger increases with the number of parameters. This is relevant to parameters in general and not just those in disease/inoculum relations in particular. In later chapters we discuss parameters estimated independently and combined to predict disease or disease progress. It is, of course, highly gratifying if disease in the field behaves as the computer predicts it should on the basis of parameters fed into it. But the problem is not only to get good agreement between prediction and reality. Possibly a more difficult problem is to determine whether any particular parameter is accurate in itself or simply counterbalanced in error by some other parameter.

Chapter 3

Effect on Disease of Variable, Limiting Factors
Other than Inoculum

3.1 THE SCOPE OF THE CHAPTER

Chapters 1 and 2 dealt with disease response to inoculum dose, all other factors such as temperature being considered equal. The amount of inoculum was the independent variable, and the other factors contributed to the parameters. Chapter 3 reverses the topic. It deals with variable factors other than the amount of inoculum, this being considered equal. The leading questions now are, if 1000 spores fall on plants and cause n lesions to form, what determines n, and, more particularly, how does n vary?

Figure 3.1 links Chapter 3 with Chapters 1 and 2. It illustrates a hypothetical situation with the number of lesions directly proportional to the number of spores, an average of five spores being needed to produce one lesion. The disease/inoculum line is straight, and passes through the origin. It conforms with Curve A of Fig. 1.7, and follows the topic of Chapters 1 and 2. The slope of the disease/inoculum line is $\frac{1}{5}$. The line makes an angle a with the x-axis, such that $\tan a = \frac{1}{5}$. This slope, or angle, is the topic of Chapter 3. It reflects the pooled contribution to disease of every factor other than inoculum.

A wide angle indicates that conditions favor disease: The host plants are susceptible; the pathogen is aggressive genetically or because of synergism within units of dispersal (as distinct from between units of dispersal, which if the synergism is facultative makes for Curve D of Fig. 1.7 without affecting the angle at the x-axis); the temperature and humidity are nearly optimal for disease; there are abundant food bases if the pathogen needs food bases; and so on. Our special concern is with how this angle is affected by variable limiting factors and interacting factors.

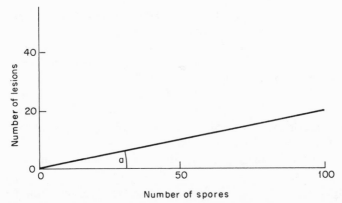

Fig. 3.1   The angle *a* measures the effect on disease of all factors other than inoculum. It measures the inoculum's potential for causing disease. Whereas Chapters 1 and 2 were concerned with the shape of the disease/inoculum curve, here a straight line, Chapter 3 is mainly concerned with what affects the curve's slope, which is tan *a*.

3.2   THE IMPORTANCE OF VARIATION

This book is concerned not so much with factors in general as with variable, limiting factors in particular. Think back to Chapters 1 and 2. In most of the graphs the amount of disease was plotted against the amount of inoculum; the graphs showed how disease varied as inoculum varied. Disease was the dependent variable and inoculum the independent variable. The key word in the previous sentence is *variable*. If inoculum were a constant factor in disease or if inoculum varied so insignificantly that it was not a significant limiting factor, Chapters 1 and 2 would not, and could not, have been written. Our topic is variation.

So too with factors other than inoculum. Their importance depends on how they vary as limiting factors. The change from Chapters 1 and 2 to Chapter 3 is a change from inoculum as the independent variable to factors other than inoculum as independent variables. Variation remains the key, both in the dependent variable—disease—and the independent variables. Consider two examples.

The effect of temperature on disease has been widely studied, and the literature of temperature effects is voluminous. Colhoun (1973) recently reviewed the literature. He placed the effects of temperature in 14 categories, e.g., the effects of temperature on spore germination. Even 14 categories are not exhaustive. Others could have been added, e.g., the effects of thermal air currents on spore dispersal. Unquestionably, temperature is a far-ranging factor in disease. But that is not to say that it is always an important variable. Royle (1973) studied hop downy mildew caused by *Pseudoperonospora humuli* in Kent. Using multiple regression equations (see Section 3.3) he found that infection was not significantly correlated with temperature. (He used mean temperatures.) The important correlations were with variables reflecting wet conditions. For the particular disease, hop downy mildew, in the particular region, Kent, in the particular years of research, 1969, 1970, and 1971, the study of mean temperature was unimportant, simply because mean temperature did not vary in such a way as to become a significant limiting factor. Some will challenge the assertion that temperature is unimportant as a factor when it is insignificantly variable and limiting. So consider another factor.

Atmospheric oxygen has not been studied as a significant factor in, say, potato blight or wheat stem rust, the two most widely studied diseases. Colhoun (1973) reviews the literature of environmental factors in plant disease without mentioning atmospheric oxygen. So too oxygen does not appear in the index of Wood's (1967) comprehensive book "Physiological Plant Pathology." In this they follow their predecessors, none of whom, to my knowledge, rated atmospheric oxygen as a significant environmental factor in foliage disease. Nevertheless it obviously is an environmental factor. *Phytophthora infestans, Puccinia graminis,* the potato, and the wheat plant are all aerobic organisms that need gaseous oxygen quite as much as *Pseudoperonospora humuli* and the hop plant need suitable temperatures. Gaseous oxygen has failed to become recognized as a significant factor in foliage disease, not because it is not a factor, but because it is not a significantly variable and limiting factor. About this, one must be consistent and avoid double standards; if atmospheric oxygen is not a factor worth contemplating because it is insignificantly variable and limiting, then temperature is not a factor worth contemplating when it is not significantly variable and limiting.

Go down to the soil, and gaseous oxygen becomes variable and limiting, and therefore a factor of demonstrable significance. *Streptomyces scabies* is highly aerobic, and loses aggressiveness when gaseous oxygen is scarce. Potato farmers with water available for irrigation can reduce scab infection by keeping soils wet at the appropriate time, thereby reducing soil aeration. On the other hand *Pythium ultimum,* notoriously destructive in wet soils, is more tolerant of poor soil aeration (Griffin, 1963a, b), and Brown and Kennedy (1966) found that preemergence killing of soybean seedlings by this fungus was increased by reducing atmospheric oxygen to 4%.

If plant pathology is to be judged primarily by its practical and economic content, the factors that matter are those that are variable and limiting, or which can be made variable and limiting. This is a hazardous guide to the choice of projects for research, because one must choose in advance of having all the relevant evidence. Nevertheless in studying a factor it is worth remembering that the variation of that factor or the scope for making it vary is likely to determine the economic use of the study. Thus, studies on the effect of root exudates on infection can profitably be coupled with studies on whether exudation varies, or how it can be made to vary, quantitatively or qualitatively.

Even within a related group of factors the scope for variation differs greatly. Among organic exogenous substrates that affect infection, the food bases in the soil that increase infection of peanuts (*Arachis hypogea*) by *Sclerotium rolfsii* vary naturally and can also be made to vary by appropriate farming practices (Boyle, 1961; Garren, 1961, 1964); pollen on leaves can be made to vary in special circumstances; but the organic debris on leaves is likely to be more difficult to manipulate, although it varies in nature.

Variable factors in infection range from the amount of sand blasting of leaves by gales to the rate of atrophy of the wing muscles of aphid vectors. The literature forms a substantial part of the literature of plant pathology itself. There appear to be few general principles about the occurrence of these factors except that the variables are multitudinous and that special detail is needed for each disease in each set of conditions. One of the variables, the susceptibility of the host plants, has Chapter 7 to itself, and two other variables, temperature and leaf-wetness, receive some attention in the present chapter. For the rest, variables are discussed only as illustrations as the occasions arise.

3.3 MULTIPLE REGRESSION ANALYSIS OF VARIABLES

Future historians may well come to regard Schrödter and Ullrich's (1965) introduction of multiple regression analysis into the epidemiology of plant disease as one of the milestones in plant pathology. This method of analysis centers around variation. It permits the contribution of each independent variable to be assessed, and gives a precision to the discussion of variable factors that was previously wanting.

In its simplest form the regression equation is

$$Y = a + b_1x_1 + b_2x_2 + b_3x_3 + \cdots$$

where Y is an appropriate measure of disease; a is the intercept; x_1, x_2, x_3, . . . are the independent variables; and b_1, b_2, b_3, . . . their partial regression coefficients. The number of independent variables is unlimited. Usually they include some appropriate measures of temperature, the duration of leaf wetness, and inoculum. The coefficients are then determined by a computer.

At all levels of the independent variables the distribution of the dependent variable Y should be normal with constant variance. When the curve of disease progress with time has the common S shape, it is appropriate to use the logit of the proportion of diseased tissue, as Schrödter and Ullrich (1965) showed. Logits were invented by Berkson (1944, 1953); the logit of y is $5 + \log_e[y/(1 - y)]$, and a table has been reproduced by Van der Plank (1963). Others have used ordinary logarithms to good effect, when they are appropriate.

A great advantage of the method is that one can use R^2, where R is the multiple correlation coefficient, to determine the contribution of an independent variable or set of variables to the total variation. Schrödter and Ullrich, working with *Phytophthora infestans* in the very susceptible potato variety Erstling, were able to ascribe about 56% of the variation of epidemic progress to the two meteorological variables, temperature and rainfall. This is a substantial percentage, especially when one takes into account that the macroclimate, not the microclimate within the crop, was being measured.

Using multiple regression analysis Dirks and Romig (1970) suggested that predictions of the cumulative numbers of uredospores of *Puccinia graminis* and *P. recondita* can be made 2 weeks before the heading stage. Further, because Burleigh *et al.* (1969) found that the number of uredospores caught on impaction traps was correlated with disease severity, disease severity could probably be substituted as the dependent variable. As independent variables they used five "biological" variables: the age of the epidemic from the date the first uredospore was trapped until the date of prediction; the rate of increase of the epidemic during two periods; the cumulative numbers of trapped uredospores at the date of prediction; and the cumulative numbers of trapped uredospores of the other rust species. They also used six "climatological" independent variables derived from records from meteorological stations of daily maximum and minimum temperatures and rainfall. Correlations were more successful with *P. recondita* than *P. graminis.*

Work on predicting epidemic development of *P. graminis* and *P. recondita* has been continued by Eversmeyer and Burleigh (1970), Burleigh *et al.* (1972a), and Eversmeyer *et al.* (1973). Inoculum has, in general, been found to have high importance relative to factors other than inoculum. For *P. recondita* this result had been foreshadowed by the work of Chester (1943, 1946). Chester found that in Oklahoma the severity of leaf rust in the ripening crop could be predicted with some accuracy from the amount of rust on 1 April, weather after this date usually having little effect.

Kerr and Shanmuganathan (1966) and Kerr and Rodrigo (1967), studied the production of spores by *Exobasidium vexans,* the cause of blister blight in tea. They used the logarithm of the number of spores per blister as the dependent variable and the logarithm of the number of blisters per 100 shoots and mean daily sunshine as independent variables.

Royle's (1973) work on hop downy mildew has already been mentioned. As one would expect with a downy mildew, wetness measured in various ways (the number of hours with a relative humidity exceeding 90%, the number of hours of leaf surface wetness with or without a rainfall contribution, and the rainfall amount) was most important.

Kranz (1968) in an unusual study analyzed the disease-progress curves of 59 fungal parasites of 42 endemic species of host plants in Guinea. Taken together, the prevalence of disease was significantly correlated with the growth stages of the hosts and the prevalence of the host species, but only insignificantly with monthly mean temperature and rainfall. In detail, the relations appeared to be specific for each host–pathogen combination.

Burleigh *et al.* (1972b) with wheat leaf rust and James *et al.* (1972) with potato blight have taken multiple regression analysis to its logical conclusion, by using it to predict losses from disease rather than disease itself. These workers are aiming straight at one of the really important targets in plant pathology.

Multiple regression analysis applied to the epidemiology of plant disease is still new. One of the present imperfections is a tendency towards overlapping and duplication of independent variables. This obscures the picture, but is excusable. The emphasis at first has necessarily been to show that multiple regression analysis can work; now that it is known to work within limits we can expect its application to be increasingly refined. With increasing refinement will come a clearer assessment of the relative importance of the different variables. There are, however, inherent difficulties of analysis, discussed in Sections 4.5 and 4.6.

3.4 THE RELATIVE IMPORTANCE OF INOCULUM AND OTHER FACTORS

The topic of this chapter is factors other than inoculum. Multiple regression studies should soon indicate for each important disease the relative importance of the various noninoculum factors, e.g., of wetness compared with temperature. The studies should also assess factors other than inoculum against inoculum itself.

There is however a more direct approach. In forecasting disease by means of multiple regression analysis a commonly used inoculum factor is the amount of disease at the date the prediction is made. Call this the initial disease. It is one of the independent variables in the analysis. Alternatively, the inoculum factor is the number of spores trapped up to the date the prediction was made. Let us think of the disease caused by these spores as the initial disease. The dependent variable is the disease we set out to forecast. Call this the final disease. Then the relative importance of factors other than inoculum can be expected to increase as the ratio (final disease)/(initial disease) increases. There is another, approximate approach to the problem. If the final disease, as a percentage, is not very high, the ratio (final disease)/(initial disease) is approximately exp(average infection rate × the units of time between initial and final assessments). In the form of infection rates and time the argument was discussed quantitatively by Van der Plank (1963, pp. 163–166) in relation to sanitation, i.e., reducing the initial disease; but *mutatis mutandis* the argument applies equally here.

3.5 TEMPERATURE AND MOISTURE

Cohen and Yarwood (1952) analyzed published data on the growth of fungi in relation to temperature. When plotted on arithmetic scales, with temperature as the abscissa and growth as the ordinate, the growth data give a modal type of curve skewed to the right. The growth rate rises more slowly as temperature increases from the minimum to the optimum than it falls as the temperature increases from the optimum to the maximum. There are approximately twice as many degrees of temperature between minimum and optimum as there are between optimum and maximum. Schrödter (1965), using Cohen and Yarwood's summarized results, found that the data could be fitted by the equation

$$Y = \sin^2(a_1x + a_2x^2 + a_3x^3)$$

Here Y is the biological activity—in Cohen and Yarwood's data it is the rate of growth of the fungus on agar in a petri dish—and

$$x = 100\ (t - t_n)/(t_x - t_n)$$

where t is the actual temperature, and t_n and t_x are, respectively, the minimum and maximum temperature for the particular biological activity of the particular organism. Schrödter determined $a_1 = 1.28$, $a_2 = -0.00746$, and $a_3 = 0.001266$, but these coefficients can be varied to suit the circumstances.

Schrödter pointed out that the mean daily temperature gives a false impression of the relation between daily temperatures and biological activity. The decisive factor is the frequency of favorable and unfavorable temperatures. Schrödter's method of measuring biological function has recently been widely used in cereal rust studies in America.

Objections to the use of mean temperatures have less force in mild, maritime climates where the daily range is small. Maximum and minimum temperatures continue to be used as variables in disease forecasts, but their value is necessarily indirect, because these temperatures are by definition temperatures of zero biological activity.

Below the minimum or above the maximum, temperatures may harm host or pathogen. Disease following freezing injury to the host is well known. High temperature adversely affects some pathogens. Hyre (1964) found that *Phytophthora phaseoli,* the cause of downy mildew of lima beans, failed to colonize if favorable periods for infection were followed by temperatures exceeding 29.5°C. A similar effect of high temperatures was reported earlier by Crosier (1934) for *P. infestans* on potatoes, but this has been disputed by Wallin and Hoyman (1958) who found that all five of their cultures in potato leaves survived a temperature of 40°C for the 22-hour duration of their experiments.

As has already been noted, Colhoun (1973) has reviewed the literature of the effect of temperature under 14 different headings. His review can be consulted for details which would be out of place here.

Moisture is one of the three factors (other than inoculum or initial disease) most commonly studied in relation to plant disease, the other two being host susceptibility and temperature. For fungi other than the powdery mildews and for bacteria, a water film is usually needed for infection. Moreover the spores of many species of fungi and the cells of many species of bacteria are often released by water and disseminated by splashing in raindrops or by carriage in water droplets. Rain and the duration of leaf wetness are therefore factors of importance in foliage disease.

Records of rainfall are usually available, but not those of the duration of leaf wetness. For the duration of leaf wetness, rainfall—was there a short sharp shower or a continuous drizzle?—and relative humidity are imperfect measures. Relative humidity is however commonly used to good purpose as an indirect measure. Thus, several methods of predicting epidemics of *Phytophthora infestans* in potatoes involve records of relative humidities above, say, 90%, although *P. infestans* will not germinate in less than a water film. In many regions, particularly those with a maritime climate, records of relative humidities are indirect guides to periods of leaf wetness, by dew or rain.

Especially, though not exclusively, in relation to humidity and wetness, the difference between the ecoclimate within the foliage and the climate reflected by standard meteorological records is important. Hirst and Stedman (1960), working with *P. infestans* on potatoes in England, inquired why there was such a dramatic change from the early phase of a seasonal epidemic with blight confined in foci to the later phase when blight spreads rapidly and the epidemic becomes general. They found no climatic changes reflected in readings from instruments in a standard screen to account for it and decided that the change came from the ecoclimate within the crop itself. Hirst (1958) had shown how an ecoclimate favorable to blight develops within the crop. At first, when the field is young, the plants small, and much soil exposed to the sun and wind, the hours of high humidity within the crop are not much greater than in a standard screen above the crop. But early in August the rows close. The foliage becomes dense and creates within itself its own humid ecoclimate largely independent of the climate above the crop. It is at this time, when the foliage closes and creates a humid ecoclimate within itself, that Hirst and Stedman found that the change occurs from focal outbreaks to a general epidemic. On this finding, a wet season is likely to be a season of blight epidemics largely because good rains make for lush foliage and the early closing of the canopy.

In published multiple regression analyses relative humidities (mostly equal to or above 90%), the duration, frequency and amount of rainfall, and the duration of free moisture on leaves have all been used as independent variables.

3.6 THE INTERACTION OF FACTORS OTHER THAN THE AMOUNT OF INOCULUM. THE INTERACTION OF THE AMOUNT OF INOCULUM WITH OTHER FACTORS

Evidence for the interaction of factors in disease has a long history. Dickson (1923) showed an interaction between temperature and the host plant. Working with the single pathogen, *Gibberella saubinetii,* and two host plants, maize and wheat, he found that low temperatures, 8°–16°C, were optimal for infection of maize seedlings but high temperatures, 16°–28°C, for infection of wheat seedlings. Dickson *et al.* (1923) explained this effect of temperature on infection. The cell walls of maize seedlings mature relatively slowly at low temperatures, and those of wheat seedlings relatively slowly at high temperatures. A similar interaction of host and temperature in disease was found by Bliss (1946), who investigated root infection by *Armillaria mellea.* The optimum temperature for root development in peach, apricot, casuarina, pepper tree, and geranium was relatively low, 10°–17°C; and the optimum temperature for infection by *A. mellea* was relatively high, 15°–25°C. The optimum temperature for root development in rose and citrus was relatively high, 17°–31°C; and the optimum temperature for infection by *A. mellea* was relatively low, 10°–18°C. Attack is favored by temperatures least favorable to the development of the host tissues that are attacked.

There are many examples of the interaction between temperature and the pathogen. Epidemics of *Puccinia striiformis* on wheat usually occur at lower temperatures than those of *P. graminis* on wheat. Epidemics of *Phytophthora infestans* on potato usually occur at lower temperatures than those of *Alternaria solani* on potato. Examples of interaction between temperature and pathogen, the host being the same, are better known than those of interaction between temperature and host, the pathogen being the same, presumably because man grows plants of economic importance over a wide range of temperatures, even outside the natural ecological range, whereas pathogens tend to remain within their ecological temperature niches.

Temperature interacts with the duration of leaf wetness in spore germination. Most spores other than those of powdery mildews need a film of water to germinate and infect; the speed of germination and infection depends on temperature; therefore the necessary duration of leaf wetness depends on temperature. Rotem *et al.* (1971) found that a 6-hour period of wetness was inadequate for the infection of potato leaves by

Phytophthora infestans when the temperature was only 5°C. Even at 10°C, which for a 24-hour period of wetness was almost the optimum temperature of infection, a 6-hour period was barely adequate. At 15°C a 6-hour period was entirely adequate. A third factor comes into this combination: host resistance. *Phytophthora infestans* takes longer to infect potato varieties of greater resistance; and Niederhauser (personal communication) found that in parts of Mexico escape from severe attack by *P. infestans* depended on the host resistance, making the time needed for infection at prevailing temperatures greater than the usual duration of leaf wetness in the field. Light is another factor that can interact with temperature. Clayton and Gaines (1945) showed that *Peronospora tabacina* on tobacco sporulates poorly at temperatures above 20°C if the light is bright, but freely at 27°C in heavy shade.

The concept of the interaction of factors is now universally accepted by plant pathologists, and it would be tedious to multiply examples. Those who want further details can profitably read Colhoun's (1973) review. But general statements about interaction are not enough. The real problem is to measure the interaction and state the result quantitatively. Possibly the main motive behind all the discussions on inoculum potenial—see the next section—has been to find a way of summarizing the effects of interaction. But the answer lies in infection rates, discussed in the next chapter. Infection rates are the resultant of the actions and interactions of all factors, identified and unidentified, of host, pathogen, and environment.

It remains to discuss the relation between the amount of inoculum and factors other than inoculum. Rotem *et al.* (1971), Colhoun (1973), and others have concluded that when conditions favor infection fewer spores suffice to bring about infection; conversely, when conditions are unfavorable to infection more spores are needed to bring about infection. The conclusion is tautological. We define conditions as favoring infection when a spore has a better chance to infect; therefore, when conditions favor infection fewer spores are needed to infect. Figure 3.1 illustrates the relation. If conditions were more favorable for infection, the angle *a* would be wider, and fewer spores on the *x*-axis would be needed for the same number of lesions on the *y*-axis.

3.7   INOCULUM. INOCULUM'S POTENTIAL. INOCULUM POTENTIAL

Inoculum was the subject of Chapters 1 and 2, in which discussion centered around disease/inoculum relations. Inoculum's potential for causing disease depends on every relevant factor other than inoculum: the host's susceptibility, the temperature, etc. In terms of Fig. 3.1, disease, the ordinate, is determined by inoculum, the abscissa, and the angle a (or, rather, the tangent of this angle) which determines the inoculum's potential.

What is inoculum potential? To many—perhaps most—of those who use the term, inoculum potential is just inoculum. To few, if any, is it just inoculum's potential. To many—and they now include the originator of the term—it combines both inoculum and inoculum's potential, that is, it includes every factor affecting the production of disease. To others it is inoculum and inoculum's potential in varying proportions, to suit their needs. To a few it is future, potential inoculum. To others it is energy of growth.

If inoculum potential is one of the most derided terms in plant pathology, it is also one of the most pervasive. It appears in literally scores of articles in reputable journals. Any future historian of twentieth century plant pathology will have to try to answer the question, what was inoculum potential? We cannot lightly dismiss it now. In any case, the very fact that it is so widely used means that, even allowing for a penchant for jargon, the term suggests some concepts that writers are trying to express.

Horsfall (1932), working on the damping-off of tomato seedlings, originated the term, and meant inoculum. A footnote said that the concept of inoculum potential carried the idea of mass action, the greater the amount of the organism present and the greater its virulence, the more severe the disease. In other words, inoculum potential carried the idea of the quantity and quality of the infectious material ordinarily called inoculum. Factors determining the inoculum's potential, the environment and the host, were ignored. From these small beginnings, a footnote to a very practical paper on seed treatment for damping-off and a concept with a very restricted scope, the term inoculum potential began. Horsfall was, of course, not the first to discuss the effect of the amount of inoculum on the amount of disease—this had been done earlier for both fungus and virus disease—but he was the first to write of inoculum as inoculum potential.

Horsfall (1938) and Dimond (1941) extended the term to include the environment. Horsfall (1945) defined inoculum potential as the equilibrium between the number of spores, number of hosts, distance between hosts, randomness of hosts, and weather factors. Dimond and Horsfall (1960) added host susceptibility to the list. In a summary Horsfall and Dimond (1963) state that inoculum potential represents the amount of inoculum and its virulence, the amount of available host tissue and its susceptibility, plus the effect of environment on all. In this final summary Horsfall and Dimond are perfectly clear. There is no ambiguity. Inoculum potential is that which produces disease and is measured by the amount of disease it produces. There is confirmation of this in details they discuss. For example, they say that what they call inoculum potential has been called the infection index by Wilhelm (1950, 1951); and when one consults Wilhelm's papers one finds that the infection index is simply the percentage of plants that becomes infected in infected soil, in other words, the amount of disease produced.

There are two difficulties. The first is that one wonders what the concept achieves. It does not help much to set out to explain the amount of disease by inoculum potential and then to evaluate inoculum potential by the amount of disease it produces. This evaluation appeared early in the literature, when Horsfall (1938) defined inoculum potential in relation to damping-off disease as "the disease-producing power" of a soil, and Dimond (1941) defined it as "the disease-producing power" of the parasite and environment; and this evaluation is consistent to the end. The naming of a quantity component in inoculum potential measured by ED_{50} does not resolve these doubts. ED_{50} can stand on its own; it is widely used without the background of inoculum potential or need for that background.

The second difficulty—the difficulty that confuses the literature—is that while Horsfall and Dimond were refining their ideas between 1932 and 1960, the term inoculum potential escaped from their laboratory into the general literature of plant pathology. Lin (1939) used inoculum potential for the number of spores in a pycnidium. They are potentially inoculum. Weihing and O'Keefe (1962) also consider inoculum potential as inoculum yet to be produced. Wilhelm (1951) is cited by Horsfall and Dimond (1963) as the first outside their laboratory to use inoculum potential in a title. Wilhelm's inoculum potential is inoculum, varied in amount by mixing infected and uninfected soils; his infection index is the percentage of plants that become infected; and the disease/inoculum relation calculated from his data is that of Curve B of Fig. 1.7. More recently inoculum potential has been defined in terms of energy, though without any reference to joules. Garrett (1970) in his book summarizes his views. He defines inoculum potential as the energy of growth of a parasite available for infection of a host, at the surface of the host organ to be infected. He applies the term not only to parasitism but also to saprophytism; inoculum potential is the energy of growth of a fungus available for colonizing a substrate at the surface of the substrate to be colonized. Garrett (1970, p. 10) stresses that "energy of growth" implies that infective hyphae are *actually growing* (his italics) out from the inoculum. So it seems that inoculum potential is a concept not applicable to virus diseases and bacterial diseases (for which we can be thankful), and that ungerminated spores are not part of the concept.

It is pretentious wordy jargon to use the term inoculum potential where plain inoculum is meant. Perhaps we ought to be grateful that pathogen potential and host potential have not yet been inflicted on us. But, in its other meanings, we should not ignore how commonly the term is used. Characteristically, Horsfall set others thinking and discussing. All factors of host, pathogen, and environment interact and are pooled in disease. Possibly the term inoculum potential so pervades the literature because plant pathologists have been groping towards some way of representing this interaction and pooling. If so, the way of inoculum potential is a blind alley. Inoculum potential cannot be quantified. True, ED_{50} has been brought into discussions of inoculum potential. But ED_{50} was introduced and is used quite independently of inoculum potential. From inoculum potential itself all one gets is the unhelpful proposition that inoculum potential determines disease, and disease measures inoculum potential.

Those given to writing about inoculum potential might try using plain words instead. Instead of writing that the inoculum potential was great, write that there was much disease or that spores were abundant or that the weather favored disease or that spores were abundant and the weather favored disease or that spores were abundant, the host plants susceptible and the weather favored disease, ... the reader will then at least know how much you know.

Infection rates supercede inoculum potential. An infection rate in a single number quantifies the whole interaction and pooling of factors. In the essential step of quantification, infection rates succeed where inoculum potential fails.

Chapter 4

Epidemics. The Time Dimension

4.1 Time as a Factor and a Dimension: How the Chapters Fit Together

This chapter is about epidemics. In part it is written to give a background to computer studies of disease progress and disease forecasting. In part it is written to fit into place in an integrated concept of plant infection.

In Chapters 1–3 time was a factor determining the amount of plant disease. But mainly it was an undisclosed factor, among the many that determine the amount of disease on some particular date or after some particular interval following inoculation. In this chapter time is changed from a factor to a dimension; and as a dimension it introduces the concept of an infection rate.

Figures 4.1 and 4.2 illustrate the change from time as a factor to time as a dimension. Figure 4.1 follows the pattern of the graphs in Chapters 1 and 2. Disease is plotted against inoculum, both on arithmetic scales. The relation shown is that of Fig. 1.1 or of Curve A of Fig. 1.7. Disease is proportional to inoculum; and the disease/inoculum relation is represented by a straight line passing through the origin. This line makes an angle a with the x-axis. With unit amount of inoculum—we can choose any amount to be the unit—the amount of disease D is tan a. Time is one of the factors determining the angle a. If disease increased with time, a would increase with time. We take the time involved in Fig. 4.1 to be the unit of time for the purpose of this figure and Fig. 4.2—it can be any number of days or years. Thus D is the amount of disease produced by unit inoculum in unit time. In Fig. 4.2 the same data are plotted differently. For simplicity we assume that disease is a "simple interest" disease (Van der Plank, 1963) so that it can

88

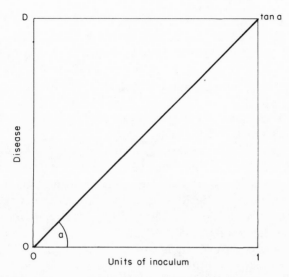

Fig. 4.1 Time is an undisclosed factor. Disease is plotted against inoculum, as in previous chapters; and the time allowed for disease to develop is one of the factors determining the angle a.

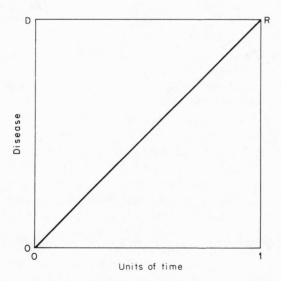

Fig. 4.2 Time is the independent variable; and the infection rate R, which is the rate per unit of time, is introduced. With other factors as in Fig. 4.1 and with unit inoculum, $R = \tan a$.

be plotted on an arithmetic scale. R is the average infection rate per unit of inoculum per unit of time, these units being the same as in Fig. 4.1. Then $D = R$, whence $R = \tan a$. From Fig. 4.1 to Fig. 4.2 the change has been from inoculum as the dimension on the x-axis to time as the dimension; and the relation between the two figures is seen to be essentially very simple.

In Fig. 4.1 the angle a (or the tangent of this angle) represents all the factors—the humidity and temperature of the air, the susceptibility of the host plants, the aggressiveness of the pathogen, etc.—that affect the ability of inoculum to cause disease. The tangent of the angle measures inoculum's potential; it pools the effect and interaction of every factor affecting disease other than inoculum. The infection rate R also reflects the angle a, as shown in the previous paragraph, and therefore also reflects inoculum's potential.

Infection rates cover everything covered by inoculum and inoculum's potential; and they cover it in a more convenient and flexible way. Some of the convenience comes from the ease with which time is measured. Disease is measured on successive dates, and the time intervals are immediately apparent. But in the long run the choice of a system of representation depends on the ability to make the system work. A system built on time as a dimension is the topic of this chapter. It provides a method for the quantification of plant disease far superior to anything that seems likely to come from quarreling over what "inoculum potential" means. But it is not to be assumed that other ways of quantification are impossible and never to be found.

It is the purpose of this book to try to give an integrated view of plant infection; and now, midway in the book, it is worth pausing to consider how the seven chapters hang together in relation to the factor or dimension of time. In the first three chapters time appears only incidentally. Chapters 4–6 are built around time. Chapter 4 and its sister Chapter 6 are about disease when time is important: epidemic disease. The amount of disease changes with time, day by day or year by year, as the case may be. Chapter 5 is the direct opposite. It is about disease when time is less important: endemic disease. Time cannot be entirely excluded as a factor or dimension; this is embodied in the principle (see Section 5.4) that a steady state of disease is impossible. But time has lost much of its importance. Chapters 4 and 5 are on different ends of a continuum, with an inevitably smudged line between. It is a measure of the continuity that the two chapters, apparently so different, are based on the same equation, Eq. (4.2), which becomes, by derivation,

Eq. (5.1) on which Chapter 5 is based. In Chapter 7, on genetic aspects of host–parasite relations, the focus is again taken off time, except, mainly, when time is needed for selection pressures to act.

4.2 THE BASIC INFECTION RATE _R_. THE PERIODS OF LATENCY AND INFECTIOUSNESS. REMOVALS

As Figs. 4.1 and 4.2 show, the infection rate _R_ links this with earlier chapters.

Recall Petersen's (1959) experiments with wheat stem rust discussed in Section 1.5 and illustrated by Fig. 1.4. Petersen allowed uredospores of _Puccinia graminis tritici_ to fall on wheat plants, and after an appropriate interval determined how much disease they had caused. At concentrations up to 2810 uredospores/cm^2 of leaf surface, disease was proportional to the number of uredospores.

Now suppose an epidemic of stem rust in wheat fields is being studied and the infection rate _R_ estimated. This rate we take as the rate of disease increase per unit quantity of sporing uredosori per day. This introduces three changes from Petersen's experiments. First, sporing uredosori instead of the uredospores themselves are taken to be the inoculum. Second, whereas in Petersen's experiments the inoculum was applied as a single dose of uredospores, not repeated, the inoculum in the field is an increasing number of sporing uredosori, added to day by day as the epidemic progresses. The epidemic situation is dynamic, not static. Third, Petersen's experiment was concerned with infection by and not the production of, uredospores; but in the epidemic both infection by uredospores to start new uredosori and the production of uredospores in uredosori are involved. In Petersen's experiments the effect of temperature, moisture, host susceptibility etc. was on the infection process alone. So too in Fig. 4.1 the angle _a_ is determined only by infectiousness and the factors that affect it. But in an epidemic, temperature, moisture, and all the other relevant factors affect _R_ both through their effects on the process of infection and on the process of spore production. If a change of temperature doubles the average proportion of spores that infect, it doubles _R_. If the change doubles the average number of spores produced by a sorus, it doubles _R_. The relation between infection, sporulation, and _R_ is as simple as that.

Put generally, _R_ is the rate of increase of disease per unit of infectious tissue per unit of time. Sporing uredosori have been used as an example of infectious tissue. The accent is on sporing and infectiousness. The infection rate R is concerned only with infectious tissue, not just infected tissue. It is concerned with uredosori that are sporing; young sori too young to start sporing and old sori that have become sterile with age are excluded. In this way R is related directly to the effect of inoculum and the production of inoculum.

Note that R assumes the disease response to be proportional to spore dose. That is, it assumes Curve A of Fig. 1.7, and none other.

The distinction between infectious and infected tissue brings in other concepts. When a lesion starts from a new infection, it is not immediately infectious. It takes time to start sporing. It passes through a period of latency p before becoming infectious. The length of this period varies greatly from disease to disease. With optimal temperatures it is about a week with stem rust of wheat, but it can be 3 years or more with a stem rust of pine trees.

Eventually, with rare exceptions, infectious tissue ceases to be infectious. The period of infectiousness i is over. Old lesions become sterile, die out or are destroyed. They are, to use the proper word, removed. With systemic virus infections removal normally occurs when the infected host plant dies. Almost the only exception to the rule that infectious tissue is eventually removed is with virus infections of vegetatively propagated plants; in so far as a clone can be considered eternal, so too is the period of infectiousness by a systemic virus it carries.

Although, with rare exceptions, removal is inevitable, it is not always great enough to be important in calculations. The more explosive an epidemic, the less the proportion of removed infected tissue; and with fast epidemics removals can often be ignored in numerical estimates without introducing substantial error. This greatly simplifies calculation.

The introduction of R, p, and i into an appropriate equation—Eq. (4.2)—is deferred until Section 4.8.

4.3   THE APPARENT INFECTION RATE r. THE INCUBATION PERIOD

The second infection rate is r. The difference between R and r is the difference between infectious tissue and infected tissue. R is the rate of increase of disease per unit of infectious tissue; r is the rate per unit of infected tissue. Thus with wheat stem rust R relates to those uredosori that are sporing, r to all uredosori including those too young yet to be sporing or even visible and those too old to be still fertile. R is intimately related to inoculum, in the sense that sporing uredosori or other infectious lesions are inoculum; r is more remotely related to inoculum, in that sporing uredosori or other infectious lesions are often only a small part of the total infected tissue.

The infection rate r cannot be considered apart from the incubation period. This is the period needed after infection for visible symptoms to appear. It is not quite the same as the period of latency. With fungus disease, lesions may become visible before they start to sporulate; that is, the period of incubation is less than the period of latency. With virus disease, plants may become infectious before they are seen to be infected.

Consider a numerical example. Suppose that the incubation period is 7 days. Disease measured on visible symptoms in a field on July 8 measures all the infection present on July 1, and disease measured on July 12 measures all the infection present on July 5. If one estimates r from the increase of visible disease between July 8 and 12 one is in fact estimating r between July 1 and 5 (assuming that the incubation period stays the same). The estimate of r is correct, but must be assigned to the appropriate dates.

An estimate of r is correctly made on visible symptoms, even though r is defined on total infection and at the times the fields are examined most infections are still invisible. The period of incubation resolves the anomaly that visible disease can be used to measure an infection rate that requires invisible to be summed with visible infections. A detailed analysis was made by Van der Plank (1963, pp. 41–43). It is required that the incubation period remains constant, or that data are corrected for a variable period.

The equation—Eq. (4.1)—defining r is given in Section 4.8.

4.4   THE TWO INFECTION RATES CONTRASTED AND COMPARED. HISTORIC (MEMORY) FACTORS

An observer of epidemics is usually interested in how visible disease progresses in the field. The rate of progress, measured by the rate of increase of visible disease, is what r estimates. From this comes much of the importance of r; from this too comes much of the relative simplicity of estimating r.

Except rarely, R is not amenable to estimation by direct observation in the field.

An estimate of r, although simpler to obtain, is inherently much more complex than one of R. Consider R first. Temperature, or any other factor except time and inoculum, affects R along only two pathways. It can affect the infectiousness of spores and it can affect sporulation (to use fungus disease for illustration). That is, temperature can affect the proportion of spores that germinate and establish new lesions; and it can affect the rate at which fertile lesions produce spores. But temperature, or any other factor except time and inoculum, affects r along four pathways. It affects r not only as it affects R but also as it affects the period of latency p and the period of infectiousness i. Each of the pathways is itself complex; and temperature for example not only acts but interacts. Thus the rate of sporulation is one of the two pathways to R and one of the four to r; and Massie *et al.* (1973) studying sporulation by *Helminthosporium maydis* on maize by multiple regression analysis found that dew temperature not only acted on sporulation but also interacted with dew period.

R has no memory of the past. It is determined by environmental and internal conditions prevailing at the time of measurement. Thus sporulation of *Helminthosporium maydis* is so directly concerned with the present that Massie *et al.* (1973) were able to ascribe 99.5% of the variance of sporulation to varying dew temperatures and dew periods, acting and interacting. But r is affected by the past as well as the present. Even events long past at the time r is measured, affect the measurement.

Multiple regression analysis is now a firmly established technique in epidemiology. So let a rule be noted. An analysis of current (present) environmental factors can explain R or sporulation or infectiousness that are the components of R. But no analysis of factors affecting r is adequate unless it includes factors preceding the date on which r began to be determined. Because r and not R is the measure of the progress of an epidemic, the progress of an epidemic after an arbitrarily chosen date can be fully analyzed only by including historic factors with the current factors.

Historic factors can act in both directions. They can increase the variance of r; in a multiple regression analysis the percentage of the variance of r that can be accounted for by current factors alone will be less than 100%. They can also decrease the variance of r; the variance will be depressed below what would have been expected from the behavior of current factors. In either direction they upset an analysis based on current factors alone.

4.5   HISTORIC FACTORS THAT INCREASE THE VARIANCE OF *r*. CRYPTIC EFFECTS OF WAVELIKE VARIATIONS. A FAILING OF MULTIPLE REGRESSION ANALYSIS AND A SIMPLE PRECAUTION

Suppose that an epidemic started from a single shower of spores arriving in the fields. Suppose further that the weather stayed the same and the susceptibility of the host plants did not change with maturity; that is, suppose that R, p, and i stayed constant. How would r behave?

Figure 4.3 illustrates the behavior of r in a fast, but not unrealistically fast, epidemic. R is taken as 10 per day, and p as 6 days. In a fast epidemic removals become relatively unimportant, and i can be ignored without incurring a significant error. At the start r fluctuates greatly—too greatly to be conveniently included in the graph. (It has a maximum value of 10 and a minimum of 0.164 per day). The fluctuations diminish with time, but even 30 days after the spore shower are still evident. (These 30 days do not include an incubation period. If r is determined on visible symptoms, this period must be added to the 30 days.)

Admittedly, Fig. 4.3. exaggerates what is likely to be found in a natural epidemic, because it assumes a start from a single spore shower. A series of showers arriving at intervals would tend to dampen the fluctuations. Also, the values of r are momentary values. In actual measurements of r, observations must be made over an interval of time, which

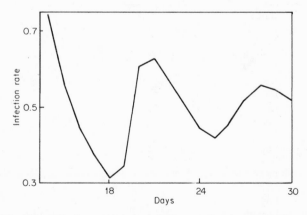

Fig. 4.3   The infection rate r in relation to the time after inoculation, e.g., by a single shower of spores in the fields. $R = 10$ per day, $p = 6$ days, and removals are negligible. (Modified from Van der Plank, 1963, Fig. 6.4, p. 64.)

also dampens the variance. Something more about this will be said soon.

In Fig. 4.3 the interval between successive crests or maxima, i.e., the wavelength, is approximately 6 days, which is the assumed period of latency p. Peterson and Jewell (1968) record that in the West of the United States before 1940, when *Cronartium ribicola* on pines was still actively epidemic, years of heavy infection were mostly at 4-year intervals, and depended in part on the buildup of inoculum in newly invaded areas on the 3- to 4-year latent period.

Table 4.1 demonstrates the wavelike variation in another way. It is assumed that an epidemic started from a single spore shower that arrived in June or July, and the equations of Van der Plank (1963, pp. 60–61) are used to calculate the multiplication of visible disease between August 1 and 8. The date of arrival of the spore shower is taken to be the only independent variable, and the chosen values of the parameters are stated in the footnote of the table. To use extremes for illustration, if the spore shower had arrived on July 2, visible disease

TABLE 4.1

The Relation between the Date of Arrival of the Spore Shower and the Estimated Multiplication of the Resulting Disease between August 1 and 8 [a]

Date of arrival	Multiplication between Aug. 1 and 8 (-fold)
June 21	9.2
June 25	6.1
June 26	5.6
June 28	6.0
June 29	8.0
June 30	10.8
July 1	15.0
July 2	16.1
July 3	13.5
July 4	11.0
July 5	9.0
July 6	6.5
July 7	4.7
July 8	3.4
July 9	3.2
July 10	4.2
July 11	7.0

[a] Estimated for $R = 6$ per day, $p = 10$ days, with an incubation period of 10 days.

would have multiplied 16.1-fold between August 1 and 8; if it had arrived a week later, on July 9, disease would have multiplied only 3.2-fold between August 1 and 8; and if it had arrived 6 days earlier, on June 26, it would also have multiplied less, 5.6-fold, between August 1 and 8. The waves are clear in the table. They oscillate about a figure of 8.1-fold, which is what the multiplication would have been between August 1 and 8 if the waves had been damped, e.g., by the much earlier arrival of the spore shower or by the arrival of a sequence of spore showers of appropriately graded size.

The arrival of a spore shower is only one of the disturbances that can lead to waves of infection. Anything that disturbs the course of an epidemic, such as a sharp drought followed by normal rainfall, will set up waves of infection. Any fluctuation of the infection rate r in the past will continue into the present, diminishing with time as it progresses.

These waves may explain the finding by Dirks and Romig (1970) in stem rust of spring wheat of a highly significant negative correlation between the slope of the regression of the daily log cumulative numbers of trapped spores on time in days up to 7 days before the date of prediction and the log cumulative numbers trapped after the date of prediction. But the explanation will be more convincing when there are data that show that the crests and troughs of the waves occurred in unequal numbers at the date of prediction.

Multiple regression analysis is not suitable for independent variables that cause disease to proceed in waves. If log or logit disease is measured at some date as the dependent variable, if the disease reaches that date in waves, and if the crests and troughs of the waves are in equal numbers, the waves will increase the variance of the dependent variable cryptically. The contribution of all the independent variables (other than the waves) to the variance of the dependent variable will add up to less than 100%. This paragraph refers to linear regression analysis currently used in the literature.

Disease developed directly from the initial inoculum—"simple interest" disease—is not in waves. But disease developed from inoculum that is itself in waves will be wavy, as when disease passes from field to field in an extended epidemic and there is an interruption or change somewhere in the epidemic during its course.

There is little in the literature of cereal rust epidemics—the most widely investigated of epidemics—that indicates how much of the variance of the infection rate can be accounted for by known factors and how much is cryptic. Burleigh *et al.* (1972a) predicted wheat leaf rust best when the predicted amount was hitched to the amount present on the date of prediction; but, more appropriate to our present inquiry, when they were unable to hitch it, they accounted for 50–60% of the variance of disease predicted 14, 21, and 30 days in advance. The independent variables used for winter wheat were the wheat growth stage, an infection function where each day is evaluated on the basis of its meteorological and biological favorability for infection, a fungal growth function based on temperature, and (in one prediction) free moisture. The variables used for spring wheat were the infection function and the minimum temperature. It is impossible at this stage to say how much of the remaining 40–50% of the variance could have been accounted for by refined equations and more data, and how much would in any case have remained cryptic. The gap must be closed, by studying interactions and variables not previously used, before we can be fully confident about the appropriateness of linear multiple regression analysis for cereal rust infection rates.

One precaution can be taken. The variance resulting from waves can be much reduced if the interval over which the multiplication of disease is observed or predicted is adjusted to at least 1.1 or $1.2p$ (Van der Plank, 1963, p. 65). In Table 4.1, p is 10 days and the chosen interval 7 days, from August 1 to 8. If the interval had been at least 11 days, from August 1 to 12, the waves would have been less conspicuous.

4.6   HISTORIC FACTORS THAT REDUCE THE VARIANCE OF r. THE PRINCIPLE OF CONTINUITY

The infection rate r is stabilized by the distribution of the lesions in age groups. If the infection rate has been high in the past, the proportion of young lesions, many of them still in the period of latency, will be high, and the proportion of old or removed lesions correspondingly low; this makes it probable that the high infection rate will continue into the future.

For illustration consider the analogy in human affairs. If the population of a country has been increasing fast, there will be a high proportion of children, adolescents and young adults, and a correspondingly low proportion of the middle aged and old. Because of the high proportion of young adults the rate of population increase is likely to be high, and it is likely to stay high as the adolescents become adult and, later, as the children become adult. Suppose now that for new and permanent economic or psychological reasons the young adults have fewer children. There will be a fall in the rate of population increase. But the age distribution of the population is still conducive to a high rate of increase—there are still many young adults—and the rate of increase will take many years before it falls to a new stable level associated with a new age grouping, with relatively less children and young adults and relatively more old people than before.

A high birth rate produces a low average age of the population; and a low average age ensures, other things being equal, a high birth rate for some time to come. So too a high infection rate produces a low average age of the population of lesions; and this in turn ensures, other things being equal, that a high infection rate will continue for some time to come.

In this analogy the average number of children born to young adults corresponds to R and the rate of population increase to r. Any change of environment affecting infection is immediately reflected in R but more slowly in r. A new stable level of r is reached only after there has been time enough for a new age distribution of lesions to be reached; and in reaching the new stable level r will vary in waves. The historic age distribution of lesions buffers future changes of r.

Fracker (1936) seems to have been the first to suggest that if infection rates start high they tend to stay high. Studying *Cronartium ribicola* mainly in the northeastern states of the United States he observed that after the fungus had developed its basic supply of aeciospore-cankers, its progress was steady and its increase inevitable, with little regard to the vagaries of humidity, rainfall, and wind velocity. The infection rate was determined by the abundance and distribution of the alternate host *Ribes* in the vicinity; with the abundance and distribution constant, the infection rate in pines stayed constant.

There is no contradiction between Peterson and Jewell's (1968) statement, quoted in the previous section, that infection by *C. ribicola* proceeded in waves corresponding to the latent period and Fracker's statement that the infection rate tends to stay constant, because they were referring to different parts of the epidemic. Peterson and Jewell specifically refer to the buildup of disease in newly invaded areas. Fracker specifically refers to disease after it has passed through its first few years of establishment and waves have overlapped and disappeared.

There is a special advantage in using a forest disease like blister blight of pines to study waves, as in the previous section, or stability, as in this section. Year-to-year progress of disease in a perennial crop is less influenced by weather than day-to-day changes of disease in an annual crop; and basic trends stand out more clearly.

Fracker also analyzed the records of Gavatt and Gill (1930) on chestnut blight caused by *Endothia parasitica,* and showed that it too proceeded in epidemics with a constant infection rate.

There is other evidence of a general tendency for fast epidemics to stay fast and slow epidemics to stay slow. There is evidence of this in the five epidemics of *Phytophthora infestans* in potatoes discussed by Van der Plank (1963, p. 29) and in the early stages of the epidemics of swollen shoot virus in cacao discussed by Van der Plank (1965). But there are also records of marked exceptions, when a change of rate followed a change of circumstances, as when infection of pines by *C. ribicola* dropped sharply after the eradication of *Ribes.* The distribution of lesions in age groups will tend to stabilize the infection rate and buffer it; but it cannot in the long run insulate an epidemic from changes in its environment.

This and the previous section discuss two sides of the same principle: the principle of continuity of epidemics. An epidemic must be judged as a whole. It cannot be considered piece by piece, interval by interval. What happens in the present is determined also by what happened in the past. A change of infection rate in the past sets up a wave that carries on, ever diminishing, into the present. But a stable rate in the past carries that stability on into the present, and helps to stabilize the rate against later changes, fast epidemics tending to stay fast, and slow epidemics slow.

The period of latency p provides most of the continuity, its contribution increasing with the infection rate. Removals, and hence the period of infectiousness $i,$ contribute relatively little to the continuity, especially when the epidemic is fast. The basic infection rate R contributes nothing directly to the continuity, but has an indirect influence by affecting r and, through $r,$ the contributions of p and $i.$

4.7 SIMULATION OF EPIDEMICS. EPIDEM AND EPIMAY. SOME CONCLUSIONS

Computers have opened the possibility of simulating epidemic sequences as they occur: sporophore formation, spore development, release, flight, landing and germination, infection, invasion of host tissue, and so on, all with their appropriate parameters and equations.

The first and still the most publicized program was EPIDEM, written by Waggoner and Horsfall (1969) in IBM 7090/7094 Fortran IV language for epidemics of *Alternaria solani* in tomatoes. The emphasis was on sporophore development. As basic material they used old records, taken mainly in 1941 and 1944, of incomplete measurements of disease progress in tomato fields. A critical analysis shows that their calculated values differ so widely from values observed in the field that details of the program need not be considered; and EPIDEM is discussed here only because the publicity given to it has misled readers into believing that the day of accurate simulation has already dawned. In reality, that day is still far off.

On the data in Waggoner and Horsfall's Table 13 the infection rate r observed in tomato fields between July 21 and September 11, 1941 was 0.058 per day. The rate calculated by EPIDEM, on the data in the top half of their Fig. 8, over the same period was 0.141 per day. Between August 8 and September 25, 1944, the observed rate was 0.069 per day and the rate calculated by EPIDEM 0.146 per day. Put differently, EPIDEM overestimated the increase of lesions between July 21 and September 11, 1941, 75 times; that is, it incurred a 75-fold error in 52 days. In 1944 it incurred a 41-fold error in 48 days. A relevant comparison of observed values and values predicted by EPIDEM in 1944 is made in Fig. 4.4 by converting logarithms into arithmetic scales. The two curves have been aligned to start on August 8 from the same level of disease—the observed level—and the great error becomes apparent. Waggoner and Horsfall did not check their numerical data, and misled themselves and their readers into believing that the epidemic had been accurately simulated. The error can be traced to a change of scales in representing logarithms in their Fig. 8. In the upper part of this figure, unit change of logarithms is represented by 7.57 mm of axis, and in the lower part by 15.1 mm. The scale is doubled; the figure erroneously makes it appear that disease trends in the upper and lower parts are similar, whereas in reality EPIDEM calculated infection rates about double those observed in the field.

In EPIMAY Waggoner *et al.* (1972) simulated southern leaf blight of maize caused by *Helminthosporium maydis*. EPIMAY was evaluated by Shaner *et al.* (1972) in Indiana, and in four of the six locations appeared to simulate accurately. But with EPIMAY too the apparent accuracy was due to numerical miscalculation. For example, at West Lafayette the simulated infection rate was stated to be $r = 0.113$ per day, which compares well with the rate $r = 0.122$ observed in the field. But

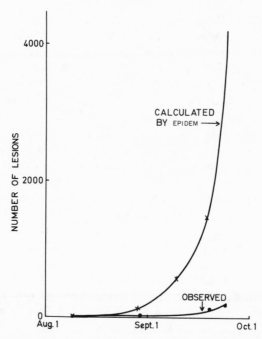

Fig. 4.4 A comparison between the increase of early blight of tomatoes observed in the field in 1944 and the increase predicted by EPIDEM. The curves have been aligned to start from the same level of disease on Aug. 8. (This figure is the delogarized version of Waggoner and Horsfall's, 1969, Fig. 8.)

a recalculation using the fitted curve in their Fig. 3 or the data in their Table 1 between the same relevant dates puts the simulated rate as about 0.24 per day, or more than double what it was stated to be. (The published details suggest that observed values of r were properly obtained by using logits, whereas the estimated values were calculated by using common instead of natural logarithms.)

Much more important in the long run is the inherent inappropriateness of EPIMAY, because infection rates were not balanced against initial inoculum. EPIMAY calculated that 1971 was more favorable to southern corn blight at West Lafayette than 1970, whereas actual disease was much more severe in 1970 than 1971. Shaner *et al.* (1972) explain this by pointing out that the initial influx of spores early in the season—the initial inoculum—was possibly 1000 times greater in 1970 than in 1971. Let us use this figure for illustration. In an epidemic lasting 60 days, a 1000-fold increase of initial inoculum is equivalent in its effect on the terminal amount of disease to an increase of 0.115 per day (more accurately, 0.11513 per day) in the infection rate. That is, a rate $r = 0.115$ per day will increase 1 to 1000 in 60 days; or, put differently, if 1000 increases to a terminal amount of, say, A in 60 days at the rate r per day, 1 will increase to A in 60 days at the rate $r +$ 0.115 per day. Now, as we have seen, the observed rate in the field at West Lafayette was about 0.122 per day. So, on the evidence available, the initial inoculum for a southern maize leaf blight epidemic is as important for the terminal amount of disease as the infection rate during the growing season, i.e., as the weather during the summer. To try to simulate an epidemic of southern maize leaf blight on summer infection alone is to misconceive the total problem from the start.

Four matters can usefully be discussed in relation to simulation.

First, simulation techniques should be married to multiple regression analysis. Multiple regression analysis is not very apt for dealing with time as a dimension, but it is probably the most useful technique for determining the parameters to put into a program of simulation. For example, Massie *et al.* (1973) have accurately determined by regression analysis how dew temperature and dew period act and interact when *Helminthosporium maydis* sporulates. Sporulation is only one of the sequence of steps in an epidemic of *H. maydis,* and equally painstaking determinations are needed for every other step. Only when this has been done can we hope to simulate an epidemic accurately.

Second, it is imperative that the strictest standards of self-criticism and reporting be maintained. The detection of error is of paramount importance, and the error must be revealed. Computers can counterbalance errors in parameters (see Section 2.10), and provide seeming agreement where agreement does not exist. Special effort, therefore, must be given to tracing error.

Third, initial inoculum needs as detailed a study as subsequent infection rates. Not only is the quantity of inoculum important, but also the timing and progress of infectiousness.

Fourth, computer programs emphasize the need for knowing classical epidemiology, even if only for heuristic purposes. Five minutes' work with a book of mathematical tables would have shown at the start that EPIMAY was unlikely to simulate epidemics of *H. maydis* adequately.

For a general discussion of methodology in epidemiological research, Zadoks' (1972) review is well worth studying.

4.8 EQUATIONS FOR INFECTION RATES. MISUSE OF THE WORD LOGISTIC

The infection rate r is defined by the equation

$$dy/dt = ry(1 - y) \qquad (4.1)$$

where y is the proportion of diseased tissue. This is a definition of r and not a description of an epidemic process. It makes r a speedometer and nothing more. This is as it should be.

There have been attempts to use the equation to describe a particular pattern of epidemics. Fracker (1936) was the first to do so. The pattern assumed is one of logistic increase. Equation (4.1) is an equation of logistic increase only if r stays constant. Admittedly there is in epidemic processes a stabilizing influence on r discussed in Section 4.6. But equally there are influences, discussed in Section 4.5, that cause r to vary; and r is sensitive to any appropriate changes in the weather and other environmental factors, as well as to changes in host susceptibility as the plants grow older. There is no reason for assuming r will stay constant throughout an epidemic, and therefore no reason for thinking of the epidemic process as logistic.

Even supposing all factors to stay constant—this is, even supposing that i, p, and R stay constant—r would not stay constant. It would be nearly constant (except for waves) until y increased to about 0.3; thereafter it would increase for a while; and finally, as removals began to dominate the epidemic, it would decrease again, dropping to zero as disease reached an asymptote. This has been discussed by Van der Plank (1963, 1968). An analysis of the literature of observed epidemics shows that the final phase of slowing down often occurs earlier than expected. Probably, the reason for this is the same as the reason for Curve B in Fig. 1.7; there is not an unlimited supply of equally susceptible sites available for infection, and the slowing down occurs when only the less susceptible sites remain available. For example, old and young leaves are seldom equally susceptible to foliage disease; and the more susceptible tend to be infected first.

The infection rate R is defined by the differential difference equation

$$dy(t)/dt = R[y(t - p) - y(t - i - p)][1 - y(t)] \qquad (4.2)$$

Here $y(t)$, $y(t - p)$, and $y(t - i - p)$ are y at times t, $t - p$, and $t - i - p$, respectively, and $y(t - p) - y(t - i - p)$ is the measure of infectious, as distinct from infected, tissue.

The rate r is measured as the regression of $\log_e[y/(1 - y)]$ on time; but only exceptionally can R be measured directly from an epidemic's progress. When there are no waves and the epidemic is proceeding at a steady rate and y is relatively small (and definitely not exceeding 0.15), R can be estimated from r by the equation

$$r = R(e^{-pr} - e^{-(i+p)r}) \qquad (4.3)$$

This is Eq. (8.5) of Van der Plank (1963, p. 102) in a more convenient form.

4.9 CONTINUOUS AND DISCONTINUOUS INFECTION. A SIMPLE RULE FOR INTERCHANGED ESTIMATES

Equation (4.1), defining r, holds for all infection, whether it be continuous or discontinuous. It has been shown by Van der Plank (1963, pp. 75–76) that r estimated as the regression of $\log_e[y/(1-y)]$ on time between dates t_1 and t_2 is the true average of all values between those dates. But Eq. (4.2), relating R, i, and p, holds only for infections proceeding continuously.

Most infection is not continuous. Many diseases, e.g., depend on dew, and infection occurs only during the time of the day when there is dew.

Corsten (1964) and Oort (1968) considered the other extreme of complete discontinuity. Infection was taken to occur only for a moment each day, at the same time each day. For the rest of the day infection ceased, to be followed by a moment of infection on the following day. Thus, a period of latency, $p = 5$ days, means that after 5 days, at the beginning of day 6, there would be infection for a moment, followed by momentary infection at the beginning of day 7, and so on for the duration of the period of infectiousness. They gave a mathematical description of discontinuous increase of disease.

Corsten and Oort misinterpreted the parameters. What they called the multiplication rate m is no other than the infection rate R first defined by Van der Plank (1963) and used in this chapter. The symbol m can therefore conveniently be dropped from future literature. But the period of latency p must be differently interpreted for discontinuous and continuous infection; if there is a case for using different symbols in discontinuous and continuous infection, the case is for p (and in certain circumstances i).

Compare for example $p = 5$ days in discontinuous infection with $p = 4.5$ days in continuous infection. As we have just seen, $p = 5$ days with discontinuous infection means that all the infection for day 6 occurs in a moment, immediately after 5 days from the initiation of the lesion. With continuous infection $p = 4.5$ days means continuous infection over the whole period from 4.5 to 5.5 days, which gives an average time of 5 days after the initiation of the lesion. That is, $p = 4.5$ days with continuous infection is approximately equivalent to $p = 5$ days with discontinuous infection. And so it is for the following 24 hours, with discontinuous infection occurring all at once after 6 days and continuous infection occurring between 5.5 and 6.5, averaging 6, days from the initiation of the lesion.

Table 4.2 puts to a numerical test the approximation that $p = 5$ days with discontinuous infection has the same effect on the infection rate r as $p = 4.5$ days with continuous infection. Four combinations of R and i are used, and the estimates of r are the settled rates after the initial waves have been damped out. For example, with $R = 2$ per day and $i = 2$ days the estimate of r for discontinuous infection with $p = 5$ days is 0.253 per day and for continuous infection with $p = 4.5$ days is 0.254 per day. The estimates agree closely. The slight overestimation of r is because continuous infection during the first half of a day contributes more to the infection rate than during the second half; a weighted value of p would be greater than an ordinary value, i.e., greater than 4.5 days. But more detailed calculation shows that the error of using $p = 4.5$ days is less than 1 hour for all four entries in

TABLE 4.2

Infection Rates for Discontinuous and Continuous Infection, with Adjustments to the Period of Latency

R (per day)	i (days)	Discontinuous infection [a]		Continuous infection [b]	
		p (days)	r (per day)	p (days)	r (per day)
2	2	5	0.253	4.5	0.254
2	5	5	0.345	4.5	0.348
5	2	5	0.423	4.5	0.424
5	5	5	0.493	4.5	0.496

[a] Calculations by Corsten's (1964) method.
[b] Calculations by Eq. (4.3).

Table 4.2; when we are able to determine p experimentally with an error of less than an hour, it will be time enough to begin thinking about a better approximation than the one we have used.

To calculate r for discontinuous infection, subtract half a day from the period of latency and calculate using Eq. (4.3), which is for continuous infection. The direct estimation of r in discontinuous infection requires a computer; the indirect estimation, via continuous infection and an adjusted value of p, takes only a few minutes and needs only a table of e^{-x}.

In Table 4.2, i is an integer and has the same meaning for discontinuous as for continuous infection. But when i includes a fraction of a day, it has a different meaning. Thus, $i = 1.1$ days with discontinuous infection is equivalent to $i = 2$ days for continuous infection. The rule for using equations for continuous infection to calculate for discontinuous infection is simple: If i for discontinuous infection is not an integer, increase it to the next integer. This rule requires that p with discontinuous infection should itself be an integer, which is necessarily so if infection occurs at the same hour each day.

4.10   THE PERIOD OF LATENCY. SOME OBJECTIONS CONSIDERED

The term, period of latency or latent period, is taken over from medical epidemiology. We shall consider first its estimation and then some objections to its use.

Estimating the period of latency p is difficult and full of snares. The period is defined as the interval needed for newly infected tissue to become infectious; in terms of fungus disease, it is the time after the initiation of a new lesion needed for spores to be released from that lesion. But there are complications hidden in this definition.

Suppose an expanding lesion was ready to start sporulating 10 days after it was established. Suppose that on day 11 it released 1 spore, on day 12 1000 spores, and on day 13 1,000,000 spores. What is the period of latency? Is it 10 days? But 1 spore is negligible compared with 1000 spores. Should we neglect the 1 spore, decide on 1000 spores, and say the period of latency is 11 days? But 1000 spores are negligible compared with 1,000,000 spores. Should we then decide that the period of latency is 12 days? Increasing sporulation by a lesion, as it expands or gets into its stride, is common, and the questions we ask are relevant. To complicate matters further, the spore on day 11 is likely to contribute more to the infection rate than any 1 spore on day 12, and any 1 spore on day 12 more than any 1 spore on day 13.

It is probably easiest to think in terms of expanding lesions such as those of potato blight; appropriate adaptations can be made for other types of lesion. Consider an expanding lesion. If a_1, a_2, . . ., a_n are the proportions of the area of a lesion with periods of latency p_1, p_2, . . ., p_n, and if sporulation is proportional to the sporulating area, the period of latency p of the lesion as a whole is given by

$$e^{-pr} = a_1 e^{-p_1 r} + a_2 e^{-p_2 r} + \cdots + a_n e^{-p_n r}$$

$$a_1 + a_2 + \cdots + a_n = 1$$

A complicating feature of this model is that in an expanding lesion p becomes a variable dependent on r. Reducing r will on its own increase p. For example, suppose a field of potatoes is sprayed with a fungicide to protect it against *Phytophthora infestans*. Suppose further that the fungicide acts purely as a protectant; it reduces the number of lesions started, but does not affect the way a lesion develops and forms spores after it has started. The lesions are fewer but normal, and each individually produces spores at times and in quantities normal in every way. Yet, because the fungicide reduces r, it increases p. The reason is this: The faster the infection rate, the more important, relatively, is that part of the lesion which starts to form spores first. In terms of potato blight, the faster the infection rate, the more important is sporulation at the center of the lesion relative to sporulation that occurs later, away from the center. Conversely, a slow infection rate, e.g., in the presence of a protectant fungicide, spreads the weight more evenly over the whole lesion and thus increases p.

A lesion normally starts at a single point of infection (except late in an epidemic when lesions crowd one another and overlap). It is the expansion of such a lesion from a single point that provides the special interest and particular mathematical difficulties of the period of latency. The experimental determination of p should therefore always be made on lesions starting at a single point. This implies using dilute inoculum to start the lesion. Concentrated inoculum, starting a lesion that is really a number of separate lesions so crowded together as to be indistinguishable as separate entities, reduces the time needed for sporulation (Lapwood and McKee, 1966; Shearer and Zadoks, 1972). Thus Lapwood and McKee found that lesions of potato blight started with 1 zoospore per drop of inoculum took a day or more longer to start sporing than those started with 256 zoospores per drop. We might hazard the guess that the reduction in time for sporulation when concentrated inoculum is used is an illustration of Klebs' (1900) theory (see Section 5.7) that in fungi vegetative growth and reproduction are antagonistic. Crowded lesions unable to expand vegetatively because they are hemmed in by their neighbors should, on this theory, begin to sporulate earlier. Be this as it may, we must regard concentrated inocula and crowded lesions as artifacts, and avoid them when determining p experimentally.

There are objections both to the concept and to the name of the period of latency.

It has been convenient to divide infected tissue into three classes: tissue still in the period of latency, infectious tissue, and removed (sterile) tissue. This brings in the concept of a period of latency extending from the time of infection to the time of infectiousness. But it would be more consistent with much of what has been discussed in this chapter if one took the period to extend from the time of inoculation that starts one generation to the time of inoculation that starts the next generation. This would mean extending the period by adding the time needed for spores to disperse, germinate, and start to infect.

The concept and name of the period of latency was taken from medical epidemiology. But latency has other and earlier meanings in plant pathology. In particular, there is a large literature of latent virus infections, meaning infections that normally cause no visible symptoms. The difficulty is that most possible alternative names have been preempted. Incubation period is commonly but wrongly used to denote the time to sporulation. The meaning given in Section 4.3, the time taken for symptoms to appear, is supported by all the dictionaries, and can be taken to be correct. (Webster's New International Dictionary defines the incubation period as the period between the infection of a plant or animal by a pathogen and the manifestation of the disease it causes. The Oxford Dictionary agrees.) Generation time, for p, has its merits because p is concerned with the interval between generations. But generation time has already been given a variety of meanings in the literature. Lapwood and McKee (1966), e.g., use generation time to mean the time for the first spores to be produced, i.e., it is the time for the generation of the first spores. This meaning has little use in epidemiology.

The period of latency, by that or any other name, needs detailed investigation, both in relation to its quantitative evaluation and its applications in epidemiology. The investigation might include the possibility of quantitative evaluation by means of the expanding lengths of infection waves in epidemics as well as by means of studying individual lesions. We can well leave to future investigators the question of whether the term, period of latency, is suitable, or whether it should be redefined, or whether it should be replaced.

Chapter 5

When Time Is Unimportant. Endemic Disease

5.1 SOME DEFINITIONS. ENDEMIC DISEASE OF PERENNIAL TISSUES. OBLIGATE PARASITES USED FOR ILLUSTRATION

The adjective, endemic, is differently applied to plants and animals, on the one hand, and their diseases, on the other. With plants and animals, endemic means native, indigenous. With diseases, it is used in opposition to epidemic. Both American and British usages agree on this.

Webster's New International Dictionary defines endemic disease as disease peculiar to a locality or region and constantly present to a greater or lesser extent in a particular place, and distinguished from epidemic or sporadic disease. The Oxford English Dictionary defines endemic disease as disease habitually prevalent in a certain country, and due to local permanent causes.

In contrast with the definitions of endemic disease, epidemic disease, according to Webster's Dictionary, affects or tends to affect many persons within a community, area, or region *at one time.* The example given is that many children died *that winter* of epidemic fevers. The Oxford English Dictionary says that epidemic disease is prevalent among a people or a community *at a special time,* and is produced by some special causes *not generally present* in the affected locality. (The italics are mine.)

The difference between endemic and epidemic disease is clear; it is about time, and not about the place where host or parasite is native. Endemic disease is constantly present, habitually prevalent. Epidemic disease is sporadic, occurring within limited time, as during an occasional winter. Nowhere in the definitions is endemic disease associated with native hosts or pathogens. There is indeed an association in that the coexistence of native hosts and pathogens tends to make disease en-

111

demic (see Sections 5.5 and 5.6), but this association is not contained in the definitions, and although it is common it is not inevitable or universal.

Time is an important variable factor or dimension in epidemic but not in endemic disease. During an explosive epidemic of stem rust in wheat, time matters greatly, in the sense that disease increases fast from day to day. Next week's disease is likely to be much greater than this week's, and it matters profoundly whether an epidemic starts 4 weeks before the fields are due to ripen, or only 2 weeks before. No reference to the amount of rust in a wheat field is useful unless it also gives the date on which the amount was determined.

Cronartium ribicola in North America illustrates both epidemic disease and endemic disease: Disease in which time is important and disease in which it is less important. When the fungus reached the New World from the Old, it caused severe epidemics of blister rust in 5-needled pines. Consider what happened in the West. For the first 15 years after it was introduced, *C. ribicola* traveled south and southeast at a rate of approximately 25 miles a year. For the next 20 years the rate fell to about 20 miles a year. Finally, during the past 20–25 years the rate has fallen to essentially zero (Petersen and Jewell, 1968). On the southern and southeastern front in the West, blister rust caused by *C. ribicola* is now essentially endemic. The sequence just described is also a sequence in the importance of time. At first, just after *C. ribicola* was introduced into the West, the disease was epidemic and time was important to the extent that the disease's front moved 25 miles in a year. Now, on the southern and southeastern front the disease is essentially endemic, and time is unimportant to the extent that the front stays more or less unchanged from year to year.

Endemic disease occurs widely. But we shall for the sake of continuity discuss it around disease of forest trees (except in the last two sections), and around diseases of tree trunks in particular. A forest environment is particularly suitable for endemic disease, and perennial organs such as trunks favor timelessness. We exploit these features for ease of illustration.

The examples used are (with the exceptions of *Fomes, Ceratocystis,* and *Endothia*) obligate parasites. This restriction was introduced in order to keep the discussion around host–pathogen interactions, without introducing saprophytic survival. In all probability conclusions apply also to the great mass of pathogens which although not obligately parasitic live a mainly parasitic existence. Our aim was to exclude from consideration those fungi that survive mainly as saprophytes, such as the slash-inhabiting fungi that are able to cause rot in hardwoods. Admittedly there is nothing in the definition of endemic disease that excludes survival as saprophytes, but emphasis on saprophytism would take us too far away from the theme of this book.

5.2   EPIDEMIC AND ENDEMIC DISEASE AS A CONTINUUM

Host, pathogen, and environment never stay in absolutely constant equilibrium. Variation is inevitable. Latitude, usually with warmer conditions at lower latitudes, longitude, if e.g., there is a gradient of rainfall from east to west, and altitude determine disease on a macroscale, to use the terminology of Van Arsdel (1965). Local topography determines disease on a mesoscale. The structure of a forest—whether e.g., the canopy has recently been opened by the death of a tree, or changed by storm damage—determines disease on a microscale. In an unworked forest, with the ages of trees in equilibrium, conditions might conceivably remain nearly constant on a macroscale or mesoscale, except for climatic changes. But on a microscale, variation is inevitable.

Because conditions cannot stay constant, at least on a microscale, it is inevitable that even in an area where disease is mainly endemic there will be local epidemics.

Consider the other extreme of areas in which disease is primarily epidemic, e.g., the Mississippi Valley wheat fields in which stem rust epidemics occur. At both ends of that wide area disease is constantly present, either on bridges of wheat or other grasses in the south or on bridges of barberry bushes in the north. And constant presence is the criterion of endemic disease.

There is no pure endemic disease on the one hand or pure epidemic disease on the other. They are mixed, and there is a continuum between the extremes. (The one exception, associated with latent virus disease, will be discussed later.)

This continuum exists even within the confines of a single disease. *Cronartium ribicola* caused severe epidemics of blister rust on 5-needled pines in North America when it reached here, and is still epidemic in many parts. But it causes essentially endemic disease in its Old World center of origin; and in the West of the United States it now has a southern front of essentially endemic disease, with little change over the past 25 years although infection continues to come and go locally.

Because of the continuum, definitions of endemic and epidemic disease are necessarily smudged. The definitions themselves must not be blamed. Endemic disease by any other name would remain in the continuum; and it will not help to cast around for other names, such as balanced disease. To substitute some other term for endemic disease would erase no smudges.

5.3   AN EQUATION FOR TIMELESS DISEASE

Except when infection is 100%, disease is not uniformly spread throughout a population of host plants. It occurs in foci, which are a topic of the next chapter. Foci are areas of higher than average disease. They may occur where conditions on a microscale (see the previous section) especially favor infection, or where inoculum happens to have landed.

Our model is a population of host plants with disease occurring in foci. Each focus starts from a small amount of inoculum, builds up, and levels off. The population of host plants is assumed to be large; and continuity of disease is given by a continuity of foci, some just beginning, some enlarging, and some leveling off with infection eventually dying out.

Disease starting from very small beginnings and increasing according to Eq. (4.2) will eventually level off at an asymptote L, where

$$L = 1 - e^{-iRL} \tag{5.1}$$

100L is the maximum percentage of disease reached in a focus as it levels off. Interpreted biologically, iR is the average number of progeny per parent infection, i.e., the average number of daughter lesions per parent lesion, when infection is unrestricted by any shortage of healthy tissue available to be infected. So too for systemic disease, iR is the average number of plants infected by a single diseased plant during its lifetime, when there is an unlimited number of plants available for infection. Table 5.1 relates 100L to selected values of iR.

Note that p, the period of latency in Eq. (4.2), disappears from the derived equation (5.1). At the asymptote L disease is timeless. Put differently, the upper level of disease proceeding according to Eq. (4.2) is determined entirely by iR, which is a number expressed without mention of units of time; and p, which must be determined in units of time, plays no part. Thus when for systemic disease we say that iR is the average number of plants infected by a single diseased plant during its

TABLE 5.1

The Upper Limit to Disease Which Starts from Very Low Levels [a] and Has Unlimited
Time to Develop

Unrestricted progeny per parent lesion [b]	Percent disease at upper limit [c]
1	0
1.01	2
1.1	18
1.2	31
1.5	58
2	80
3	94
4	98
5	99.3 [d]
∞	100

[a] Disease near the limit of zero. For this reason the first entry in the second column is 0.

[b] This is iR of Eq. (5.1); it is the average number of lesions produced during its whole lifetime by one parent lesion, when there is unlimited healthy tissue available for infection.

[c] This is $100L$ of Eq. (5.1).

[d] An illustration of the integration of Eq. (4.2) with $iR = 5$ and $L = 0.993$ was given by Van der Plank (1965, Fig. 3, Curve C). The value of p used in the illustration is irrelevant to our present discussion, because it does not affect the value $L = 0.993$, but only the rate at which it is approached.

whole lifetime, we are not concerned in Eq. (5.1) with when during that lifetime infection was passed on; it is immaterial whether it was passed on early or late.

Consider Table 5.1 briefly. Unless each parent lesion leaves more than one daughter lesion, i.e., unless $iR > 1$, disease will not occur. This common-sense threshold theorem was stated earlier by Van der Plank (1963, p.102). If under unrestricted conditions each parent lesion leaves an average of 1.01 daughter lesions, i.e., if $iR = 1.01$, disease will take off and finally level out when 2% of the susceptible tissue has been infected. Put in terms of a systemic disease of trees, if under unrestricted conditions for spread 100 infected trees during their lifetime cause the infection of an average of 101 other trees, disease will level out when 2% of the population of trees have been infected. If $iR = 1.1$, disease will level out at 18%. And so on. Finally, if $iR = \infty$, disease will continue until infection is complete; this commonly happens when host plants carry a virus latently—tristeza disease of orange trees on tolerant rootstocks is a familiar example.

5.4   LOSS OF INFECTIOUSNESS. ENDEMIC DISEASE LIKELY TO BE
UNDERESTIMATED. A STEADY STATE IMPOSSIBLE WITH
OBLIGATE PARASITES. ANTI-EPIDEMICS. LATENT
VIRUSES AN EXCEPTION

The limit L is reached when all diseased tissue has ceased to be
infectious. The blisters, conks, cankers, and galls are all old and sterile,
or dead or perhaps eaten by insects, birds, or rodents. Diseased trees or
branches produce no more spores.

A forest consumes its dead. Saprophytes invade old, sterile diseased
tissue, and rot it. Animals invade this tissue, and eat it. Neighboring
trees overgrow trees that have died. Sooner or later there is little evidence that disease once occurred, except debris and scars here and
there.

In this there is a sharp difference between epidemic and endemic disease. In epidemic disease, lesions are usually fresh and countable. One
can walk into a field of wheat and estimate the amount of stem rust. In
endemic disease, lesions are usually old and often unrecognizable, and
uncountable. Thus, we may use Table 5.1 to estimate that if with unrestricted spread each parent lesion leaves an average of 1.1 daughter lesions, disease will level off at 18%. But it is unlikely that that 18%
disease will be seen. Most of it may be beyond diagnosis, or the diseased tissues may be consumed. The nearer endemic disease is to timelessness, the more it will be underestimated, through lack of evidence
that it existed.

The previous paragraph is not concerned with estimation of economic loss. Economic loss depends on relevant detail. If a young tree is
killed by disease and its place taken by its neighbors, the economic
loss may be small. On the other hand, a small scar of disease in a valuable piece of lumber destined for the sawmill could cause disproportionately large loss.

Variation in the conditions affecting disease is necessary for the survival of obligate parasites. This is a direct consequence of the loss of
infectiousness that accompanies the reaching of the limit L. If infection
were to proceed in host plants genetically and environmentally uniform,
an obligate parasite would reach the limit and die out through lack of
infectiousness. Survival of obligate parasites depends not only on conditions being favorable to infection but also on their being variable. The
greater and more frequent the variation, the more the parasite benefits.
It also follows that with obligate parasites absolutely timeless disease is

impossible; there are only degrees of importance of time. The development of disease in foci, the model for Section 5.3 and the topic of Chapter 6, is not only a matter of general observation, but for obligate parasites in particular also one of theoretical necessity.

What has just been written can be put as a general principle: With obligate parasites at least, there can be no steady state of disease. Disease must always increase or decrease. When disease occurs over the whole available area, as on an island, the level of disease is always rising or falling. When it occurs over only part of the geographical area not only does the level of disease rise and fall, but the boundary of the diseased area advances and recedes. Either advancing or receding, disease at the boundary does not meet the requirement of constant presence in the definition of endemic disease. An advancing boundary means that there is an epidemic there. A receding boundary falls outside the definition of both epidemic and endemic disease. A receding boundary means that there is an anti-epidemic there. An antiepidemic, which is the sporadic absence of disease, means that $iR < 1$ just outside the receding boundary. For obligate parasites, an area of endemic disease that does not occupy a whole geographical area is bounded by a zone of alternating epidemics and antiepidemics, as iR at the boundary is alternately greater than and less than 1. From this it follows that antiepidemics are a phenomenon of geographical areas that have disease boundaries within the area, i.e., there are no antiepidemics if the disease permanently extends over the whole area, as it might on a small island. What holds for obligate parasites probably holds for all pathogens in which a parasitic phase is essential to survival.

Perhaps the fullest information about endemic disease is for the non-obligate parasite *Ceratocystis fagacearum* that causes a systemic wilt disease of oaks in the United States. In the early part of this century it seems to have occurred with a low incidence. Neither Boyce (1938) nor Baxter (1943) who wrote comprehensive books on forest diseases in the United States mentioned it, although it had almost certainly been observed before then. Henry *et al.* (1944) review the scant earlier literature of oak disease that might have been oak wilt. They also isolated the pathogen and showed that the disease was infectious. Thereafter, and especially in the 1950's, the literature became extensive, reflecting the contemporary prevalence of the disease; and Hepting (1971) was able to say that the literature of oak wilt surpasses in volume that of any other tree disease with the possible exception of white pine blister rust. To turn to the Appalachian area in particular, West

Virginia was near the center of the endemic area which had a northern boundary in Pennsylvania. Typical of endemic disease, this northern boundary, though it must have fluctuated, showed no progressive and maintained advance; and Merrill (1967) records that in the 15 years of observation prior to 1967 the disease did not spread out of the south-western counties of Pennsylvania. In retrospect it can be seen that there was an epidemic in the Appalachian area in the 1950's and early 1960's. By the 1960's the epidemic was beginning to abate. In West Virginia the infection rate r fell steadily from 0.7 per year from 1955 to 1956 to 0.3 per year from 1964 to 1965 (Merrill, 1967). In Pennsylvania the rate was slower, about 0.1 per year. Then in 1967 disease leveled out, i.e., it had reached its asymptote; and this was followed by an antiepidemic with disease falling to negligible amounts (Merrill, personal communication 1974).

Oak wilt in the Appalachians gives a clear picture of the unsteady state of a disease that is essentially endemic. Antiepidemic follows epidemic. But there is much that is still obscure about the process. It is reasonably clear why an antiepidemic should follow an epidemic; the amount of infectious tissue drops as the asymptote is approached. But what starts an epidemic phase is not clear. Is there a builtin cyclic pattern not yet revealed? Or are there periodic environmental changes? Survival from an antiepidemic to the next epidemic may well depend on the host range. The host range of oak wilt includes all *Quercus* spp. that have been tested as well as some near relatives in *Castanea, Castanopsis,* and *Lithocarpus.* What is probably important is the wide range of susceptibility, the red oaks being more susceptible than the white. What is perhaps even more important is that members of the white oak group die slowly and sometimes harbor the pathogen in the vascular system without showing symptoms. Tolerance by the host allows a long period of infectiousness i and hence a greater chance of survival by the pathogen.

A large area of endemic disease within a continent is likely to be protean in shape, bulging here and shrinking there in a continuous flux. With changing conditions along some parts of the boundary iR is likely to be on the increase and exceed 1, while along other parts it may well be on the decrease and fall below 1. This seems to be the picture now with oak wilt in North America. The disease is apparently retreating in southern Pennsylvania but extending into North Carolina where it has been picked up in the piedmont over 150 miles from where it was prevalent before; and in northern Minnesota it has flared up in an infection center where it was previously scarcer (Merrill, 1974, in a personal communication). The oak wilt fungus has an indigenous insect vector, the oak bark beetle, and it is possible that changes in the distribution of oak wilt have been accentuated by predator–prey or parasite–host cycles. But protean changes in the boundary of large areas of endemic disease are likely to occur even with diseases not transmitted by vectors, the cycles of which are not necessary for the concept of the fluctuation of endemic disease.

There is one exception to the rule that the amount of disease must vary if obligate parasites are to survive, and that with obligate parasites absolutely timeless disease is impossible. The exception is when the parasite is carried by the host without significant harm, as when trees are 100% infected with a latent virus. No variation in the amount of disease is then possible, and disease is timeless. In Eq. (5.1) the asymptote $L = 1$ is unique. There is an absolute discontinuity between $L = 1$ and $L < 1$. When $L < 1$, it is approached as all infectiousness is lost. This has been the theme of this section. But $L = 1$ implies no loss of infectiousness, i.e., no removals. This in turn implies that constantly present infection is not harmful enough to reduce the hosts plants' fitness and endanger their survival.

5.5 ENDEMIC DISEASE AND NATIVE DISEASE. SPORADIC EPIDEMICS OF DISEASE USUALLY ENDEMIC

Endemic disease is not necessarily disease native to the country. The dictionaries agree about that. Nevertheless, there is a loose biological connection, in that agelong coexistence of host and pathogen in their native habitat is likely to bring about a state of almost, but not quite, timeless balance. There are however many exceptions in both directions, with foreign hosts or foreign pathogens living in a state of endemic disease, or with epidemics flaring up even when both host and pathogen are native. This section is about epidemic disease with native hosts and pathogens.

Epidemics can start when the old balance between host and pathogen is disturbed by a permanent change in the environment. Bingham *et al.* (1971) review the literature about fusiform rust of pine trees caused by *Cronartium fusiforme* in the southern United States. Before 1930 the disease was rare. Now it is epidemic locally, especially on *Pinus taeda* and *P. elliottii*. The environmental changes that caused this change in disease are probably changes in the relative proportions of the tree species and fertilization of the soil.

Fluctuations in the environment, as might be caused by fluctuating weather, cause sporadic epidemics. Considered generally, all native American hard pines in their natural environment seem to have enough resistance to native stem rust fungi to keep disease at low levels, but Peterson and Jewell (1968) and Bingham *et al.* (1971) describe occasional outbreaks. These outbreaks often involve a burst of infection during a single season or even a single moist period, and are then followed by years of tapering off as infection dies out. Longer epidemics also occur. Canker dating has given information about the life history of one epidemic. An epidemic of *Cronartium comandrae* on *Pinus contorta* occurred between 1910 and 1945 in the Rocky Mountain region, but disease has now returned to its old endemic level. Epidemics of native rusts are usually infrequent. In the West major outbreaks of *C. occidentale* and *C. comandrae* come decades apart. Even the pine-to-pine infecting fungi, *Peridermium harknessii* and *P. filamentosum,* despite their simplified life cycle without telial hosts, cause abundant infection only occasionally, as in one year among many. The extent of the epidemic varies. In the central Rocky Mountain region wave years of *P. harknessii* can be identified over only a few miles; whereas in the Northwest wave years affect larger areas.

From what was said in the previous section, these fluctuations are needed for the fungi to survive, and on the evidence must be judged to be sufficient. The assocation between the North American hard pines and the native stem rust fungi illustrates the compromise between the stability expected from the ancient coexistence of host and parasite and the instability without which the destructive obligate parasites could not survive.

5.6   THE IMPLICATIONS OF ENDEMICITY. HORIZONTAL RESISTANCE IN THE HOST PLANTS. THREE PROPOSITIONS

Endemicity implies both balance and coexistence. The average number of progeny per parent lesion is balanced about one, when tissue available for infection is unlimited; and coexistence of host and parasite reflects the constant presence of both that is the essence of the definition of endemic disease.

Balance implies that the resistance of the host plants is adequate under the conditions of the environment to keep the average number of unrestricted progeny per parent lesion down to about one. With less resistance the number would have exceeded one, and an epidemic would have followed.

Coexistence implies that the pathogen does not die out. It would die out if the average number of unrestricted progeny per parent lesion were consistently less than one. It may be less than one over part of the range of the endemic disease, but it cannot be less than one over the whole range. The resistance of the host where disease is endemic is finely balanced, with enough to curb the pathogen but not enough to drive it to extinction. This resistance is of course related to the environment as well, less resistance being needed to bring about a balance when the environment is less favorable to disease.

Endemicity has a clear, if perhaps unexpected, message. Unnecessary, i.e., excess, resistance is harmful to the host plants that have it. (This matter is taken up again in Section 7.18 where excess resistance ·is defined.) Resistance is not increased beyond the point of balance between host and pathogen that is needed for endemic disease.

The same message, that unnecessary resistance harms the host, comes through from an allied source. When a pathogen is taken from its center of origin to another continent climatically favorable to it, as when *Cronartium ribicola* was brought from the Old World to the white pines of the northern United States, it commonly becomes epidemic. The inference is that resistance in the host plants was low because, until the pathogen reached them, resistance was unnecessary.

The resistance involved is horizontal resistance; and the proposition that unnecessary horizontal resistance harms the host can easily be understood only if we accept two other propositions also discussed in Chapter 7.

The first of these other propositions is that horizontal resistance is not governed by special resistance genes, but by genes that have normal functions in the uninfected host plants (see Section 7.15). It will be suggested in Chapter 7 that increased horizontal resistance in outbreeding plants might be acquired by an increase in homozygosity, and this could harm the host when it is unnecessary. More generally, one might assume that the optimal balance of metabolic processes in uninfected plants is not necessarily the optimal balance in plants threatened by disease, from which it would automatically follow that unnecessary horizontal resistance would be disadvantageous.

The second of these other propositions is that horizontal resistance is not the sort of resistance that primarily decides the range of hosts of a pathogen. The host range is determined by immunity. Immune plants are nonhosts, by definition. Immunity, it will be shown, is related to vertical, not horizontal, resistance (see Section 7.19). Thus when *Cronartium ribicola* reached North America, immunity processes related to vertical resistance determined which species of *Pinus*—*P. strobus, P. monticola, P. lambertiana, P. flexilis,* and *P. albicaulis*—it would attack. But within this host range the resistance that determined the intensity of attack was probably largely horizontal. (Horizontal resistance can however determine "population immunity" as distinct from plant immunity. See Section 5.10. To this extent horizontal resistance can determine host range within a particular environment, whereas true plant immunity operates irrespective of environment.)

Bingham *et a.* (1971) have made the coexistence of host and pathogen the substance of an interesting philosophy for breeding forest trees for resistance to plant disease.

5.7 ADAPTATION IN THE PATHOGEN TO ENDEMIC DISEASE

In epidemic disease, time is important. The pathogen must hurry to attack while it can. An obvious adaptation is a short generation time; and Van der Plank (1963, pp. 52–53, 107–108) has shown that, at the extreme, the maximum infection rate is determined by the period of latency alone. The shorter the period the faster the rate.

In endemic disease, time is unimportant. Equation (5.1) shows that, at the extreme, the period of latency p ceases to be a factor in the survival of the pathogen. At the extreme, survival is determined by $iR,$ the unrestricted number of daughter lesions per parent lesion.

Neither extreme, of an explosive epidemic determined by p alone or timeless survival determined by iR alone, will be found in the continuum discussed in Section 5.2. But the extremes illustrate the importance of p in epidemic disease and the importance of iR in endemic disease.

It is a feature of many pathogenic fungi in forests that they have long periods of latency and abundant sporulation. Periods of latency of 3 or 4 years are not uncommon, as in some stem rust fungi in pines. Abundant sporulation, as when a single perennial sporophore of *Fomes applanatus* produces 30 billion spores daily measured over a period of 6 months (White, 1919), is also a common feature of many pathogenic fungi in forests. These two features adapt the fungi to survival in endemic disease. In saying this, it is not asserted that endemicity is the one and only factor to have exerted selection pressure. Many other factors may be involved. Thus, a woody substrate has made long-lived sporophores possible. All the factors interact to determine the reproductive patterns of forest pathogens; these patterns are among the most interesting ecologically in the whole of botanic science.

To survive, *F. applanatus* must achieve a daughter infection/parent infection ratio slightly exceeding 1. If an average parent infection produces thousands of billions of spores, we can safely infer that anything considerably less would fail to produce a daughter lesion, and the fungus would not survive. The vast production of spores is a measure of the resistance of bark-armored trees to infection.

It seems likely that the fungus pathogen is faced with a choice of alternatives. Either it can have a short period of latency or it can have the ability to produce billions of spores from a single lesion. It seems that it cannot have both. From research carried out last century Klebs (1900) concluded that there was a conflict between active vegetative growth and the initiation of reproductive activity in fungi, and Hawker (1957) has reviewed other evidence. To be able to bear the great mass of reproductive tissue needed to produce spores by the billion the fungus must first establish a mass of vegetative mycelium. The greater the mass, the longer it will take to accumulate it, and hence the longer the period of latency will be. Conversely, if a sorus of *Puccinia graminis* is to start producing uredospores a week after its initiation, the mass of reproductive tissue available is necessarily limited, and (compared with *F. applanatus*) so is the number of spores per single sorus.

It seems reasonable to argue in sequence that a tree trunk is hard to infect, therefore the probability of a single spore infecting it is very small, therefore each fungus colony must produce on an average billions of spores to ensure survival, therefore the production of spores must be delayed until the colony has an adequate vegetative base, therefore the period of latency of the disease is long, which is no great disadvantage to the pathogen in endemic disease for reasons obvious from Eq. (5.1). The need for coexistence of host and parasite in the long run, which is the need for a balance between parasitic aggressiveness, host resistance, and environment, will determine average spore production and the linked period of latency.

5.8 THE RARITY OF HARMFUL VIRUS INFECTIONS IN FOREST TREES

Virus disease seldom harms forest trees conspicuously. Only 0.7% of Hepting's (1971) book "Diseases of Forest and Shade Trees of the United States" is about virus disease. If one excludes the genera primarily grown as horticultural crops (*Aleurites, Citrus, Ficus, Juglans, Malus, Persea, Prunus,* and *Pyrus*), the figure drops to about 0.4%. All this contrasts strikingly with the literature of plants in general; more than 20% of the literature of plant pathology is about viruses.

The small percentage for viruses in forest trees seems to be unchallengable in substance. Other books than Hepting's on forest pathology tell the same story. Hepting's book is authoritative. It is an Agriculture Handbook of the United States Department of Agriculture Forest Service, with 1669 references to literature. The author states that the manuscript in mimeographed form was distributed to more than 300 pathologists, foresters, and others, and that the additions they suggested have been incorporated in the final version. Conspicuous harm over the years by a disease of any sort would not have been overlooked and unrecorded in the book.

The same two virus diseases—phloem necrosis of elm and witches broom of *Robinia*—which receive most attention in the book have been known for a long time; and no new and conspicuously harmful virus infections have been added to the list of forest diseases in recent years.

There is only one record, on *Picea excelsa* in Czechoslovakia (Čech *et al.,* 1961), of a virus disease of a tree belonging to the gymnosperms.

The absence of disease in vulnerable host plants is as significant as its presence. The presence of disease implies the susceptibility of the host plant. The absence implies immunity, resistance, or tolerance in the population (see Section 5.10). In the short run, the study of disease and therefore susceptibility is of obvious importance. In the long run, the study of the absence of disease is of equal or greater importance, because it is the study of population immunity, resistance, or tolerance, which are the aptest weapons for disease control. Studying resistance over narrow ranges, as when one variety of wheat is resistant to stem rust and another is not, is all very well, but plant pathologists may be doing their science a disservice by ignoring wider phenomena.

In the preceding paragraph the word, disease, was used with its literal meaning of dis-ease. According to Webster, plant disease is an impairment of the normal state of the plant body or any of its components. Harm is implied, and escape from disease implies tolerance to disease as well as resistance to it. The term, harmful disease, is a tautology; it is better to write of harmful infection, when harmfulness is to be emphasized. This is not discussed out of pedantry, but to stress that in any attempt to explain the scarcity of virus disease in forests one must impartially consider immunity, resistance, and tolerance.

Among angiosperms at any rate, there seem to be no taxonomic reasons for the scarcity of virus diseases in forests. To consider one example, tomato spotted wilt virus attacks a vast range of dicotyledons and several monocotyledons, but no forest tree diseases caused by this virus are known.

Nor does the tree form explain the scarcity of virus diseases. To refer to tomato spotted wilt virus again, the thrips vectors have a prediliction for feeding in flowers, so bark and wood are no barriers to infection. In any case, deciduous trees put on an annual flush of tender new growth comparable with that of any annual herb.

Both the longevity and size of trees could be expected to make them vulnerable to systemic infection if they were susceptible. A tree that puts out an annual flush of growth for 100 years has 100 times as long to become infected systemically as an annual plant. A tree that puts out 100 new shoots a year is 100 times as large a target for systemic infection as an annual plant making only one shoot. Escape, in contrast with immunity, resistance, or tolerance, can be ignored as an explanation for the scarcity of virus disease in forest trees as compared with annual plants.

Immunity—the presence of nonhosts—almost certainly plays a part, as it does in herbaceous host plants. Thus, Posnette *et al.* (1950) found cacao virus 1*A* to attack members of the Bombacaceae and Sterculiaceae but not eight other families tested, which included many trees.

Resistance, divorced from tolerance, would be expected to manifest itself in a low level of harmful infections, with host and virus coexisting. On an average there would be not much more than 1 daughter infected plant per parent infected plant. This sort of endemic coexistence, typical among obligate parasites like the endemic pine stem rusts, the mistletoes, and the dwarf mistletoes, does not appear to have been described for viruses in forest trees. One is left in doubt about the existence of a narrow balance between resistance and susceptibility within the local environment. As was noted in Section 5.6 the balance requires unnecessary resistance to be disadvantageous to the host plants. Without this disadvantage the average number of daughter infections per parent infection could fall well below 1, and infection would disappear. Almost all the evidence that unnecessary horizontal resistance can be disadvantageous to the host plants is for fungus disease. Virus diseases may be different; and it could well be that the difference in metabolism between resistant and susceptible host plants might be less when the pathogen is metabolically so relatively simple as a virus. The matter does not seem economically very important for virus diseases in forests, but needs investigation in relation to virus diseases of field crops.

Tolerance seems to be a widely used mechanism. It is best known from trees used in horticulture. Orange trees are tolerant of three very common viruses—those of tristeza, exocortis, and xyloporosis—that came into prominence only because horticultural practice caused orange trees to be grown on rootstocks of other species. When orange trees are grown from seed, as forest trees are, they flourish without symptoms of virus disease even in areas where tristeza virus and its aphid vectors are common and trees quickly infected.

Except for phloem necrosis of elms, the known viruses of forest or shade trees of the United States are carried by the hosts with substantial tolerance. Witches' broom, the virus disease of *Robinia* and *Gleditsia,* probably ranks next to elm phloem necrosis. But it is largely a disease of areas in which the locust trees have been felled and allowed to sprout again. The greater intolerance of coppice shoots, or of the side shoots of herbaceous plants that have been cut back, is a well-known phenomenon of virus diseases. This virus disease and others reported from forest trees in the United States and discussed by Hepting (1971) —a witches' broom of *Albizia* resembling that of *Robinia,* a line pattern disease of the foliage of *Betula lutea,* isolated chlorotic islands which are produced during the middle of the growing season in leaves of *Celtis occidentalis* and may be due not to a virus but to direct damage by the feeding of leafhoppers, a ringspot disease of the foliage of *Fraxinus americana,* a witches' broom disease of *Ostrya virginiana* of unknown cause but possibly a virus disease, and a leaf mosaic (in New York) and a yellows disease (in Texas) of *Sassafras albidum*—might well be indications of a great array of undetected, tolerated viruses in forests.

The work of Posnette *et al.* (1950) in West Africa reveals something about the mechanism of tolerance. Cacao (*Theobroma cacao*) is a New World species that was taken across the Atlantic and grown in West Africa. Here it became seriously attacked by several viruses. Posnette *et al.* showed that the source of the viruses was the indigenous forest, and

that the viruses attacked three species of the Bombacaceae and four of
the related Sterculiaceae. But for the introduction of cacao the presence
of these viruses in the native trees would have been undetected and un-
suspected. Symptoms in these native trees when artificially infected
tended to occur only when they were newly infected and disappear later
when the virus concentration seemed to become very low. The shock
phase of the disease was succeeded by a chronic phase of apparent
harmlessness.

5.9 THE CONSTANT ABSENCE OF DISEASE

Endemic disease is disease constantly present. Equally worthy of
study is disease constantly absent, even when host and parasite coexist.

Viruses provide a noteworthy example. There is no record of any
harmful virus infection of any ecologically dominant perennial plants
growing in their natural habitat. Viruses strike elsewhere—in crop
plants and orchards, in gardens, in weeds, in annual and biennial
plants, in undominant shrubs and undergrowth. But they harm neither
the dominant trees of natural forest nor the dominant perennial grasses
of natural grasslands. Reference to any book on plant virus diseases
will confirm that.

In this, viruses differ from fungi; the freedom of ecologically dominant perennials from infection is a feature of viruses in particular, not
pathogens in general.

As an example consider the streak virus of maize and sugarcane. It
is native to Africa; and in its various strains it attacks many species of
grasses. McClean (1947) has listed 24 grass hosts. The outstanding impression one gets from the list is the disproportionately large number of
annuals and short-lived perennials: cereals and weeds like *Digitaria
horizontalis*. Of the uncultivated species three-fourths are weeds in cultivated fields. Admittedly this proportion is probably somewhat biased,
because many of the records are from observations in or near fields of
infected maize or sugarcane. But this just emphasizes the point that the
virus spreads from infected maize or sugarcane to attack many short-
lived grass weeds near the fields, but entirely fails to establish itself in
the great stretches of ecologically dominant grasslands surrounding the
fields. These ecologically dominant perennial grasses are there in countless numbers year after year, potentially very vulnerable to systemic disease because of their longevity and abundance but nevertheless remain-

ing unharmed. One of these ecologically dominant grasses *Eragrostis curvula* has indeed been artificially infected in a greenhouse by Gorter (1953) using infective leaf-hoppers taken from infected maize plants. But no infected plant of *E. curvula* has ever been found out of doors. The virus clearly cannot establish itself either endemically or epidemically in *E. curvula*. In symbols, iR is substantially less than 1.

Taxonomic affinity within the grasses plays little part in determining which species become infected with streak virus. Near infected maize and sugarcane fields the common perennial climax grasses (such as *Themeda, Cymbopogon,* and *Hyparrhenia*) are, like maize and sugarcane, andropogonoid, but remain uninfected. In contrast, of the species found by McClean (1947) to be naturally infected only three are andropogonoids, the rest being festucoid (1), cloridoid (10), and panicoid (10).

Sugarcane mosaic virus and maize dwarf mosaic virus are other viruses in which ecological status and growth form overshadow taxonomic affinity in determining host ranges within the Gramineae. These two aphid-borne viruses are closely related or identical. Soon after sugarcane mosaic virus was found in the United States Brandes and Klaphaak (1923) reported that surveys in Louisiana, Georgia, and Florida had revealed infection (apart from in sugarcane) only in six species, all of them annuals: sorghum (an andropogonoid), pearl millet (a panicoid), and four wild grasses (all panicoids). To jump nearly 50 years of history, Ford and Tosic (1972) after 5 years of research in Iowa recorded natural infection by maize dwarf mosaic virus strain A in only *Sorghum halepense* (Johnsongrass, a perennial) and "some annual grasses," and natural infection by strain B only in maize. In Europe maize mosaic virus and sorghum red stripe virus, which seem closely related to sugarcane mosaic virus, have only *S. halepense* as a natural source (Ford and Tosic, 1972). Take away cultivated crops and annuals, and there is little or nothing left of the natural host range despite the vast array of perennial species in the Gramineae. (Ford and Tosic record a long list of host plants susceptible to maize dwarf mosaic virus by mechanical inoculation with sap extracts and an abrasive. These are artifacts irrelevant to our particular discussion; but they at least serve to support the proposition, discussed in the next section, that the ability of an individual plant to harbor a virus does not necessarily imply the susceptibility of the plant population to natural infection by that virus.)

One assumes that the resistance or tolerance of ecologically dominant perennial plants to viruses has been developed by selection pressure. If these plants had been susceptible, their longevity and abundance would have rendered them very vulnerable to disease, especially systemic disease; and severe disease would have destroyed the fitness that dominance implies.

The same argument about selection pressure towards resistance applies to fungus disease as well, but there are differences. Perennials are especially vulnerable to systemic infection which, once acquired, persists until the host dies. Systemic disease is characteristic of viruses, but rarer with fungi. (In the statistical record of plant diseases there is some evidence that the systemic wilt diseases caused by fungi are relatively less common in ecologically dominant perennial plants, oak wilt caused by *Ceratocystis fagacearum* being one of the few exceptions.) Second, to repeat a surmise in Section 5.6, the disadvantage to the host plants of unnecessary horizontal resistance may be less, or absent, when the pathogen is a metabolically relatively simple virus. Granted this, it would be easier for selection pressure to increase resistance to viruses beyond the point where disease in host populations is possible. The argument is stated more fully in the next section.

5.10 SUMMARY: EPIDEMIC DISEASE, ENDEMIC DISEASE, AND NO DISEASE. HOST POPULATION IMMUNITY AND RESISTANCE DISTINCT FROM INDIVIDUAL HOST PLANT IMMUNITY AND RESISTANCE

For the purpose of this summary we shall not discuss tolerance in the host and latent infection.

In epidemic disease the average number of daughter infections per parent infection exceeds 1. In endemic disease the number is not far from 1, being greater than 1 on an average but often less than 1 locally and temporarily. When the number is consistently less than 1 there is no disease. The numbers here are for unrestricted spread of infection when the amount of susceptible tissue still available for infection is unlimited.

When the number is considerably greater than 1 disease in time will reach destructive levels. This happened when *Endothia parasitica* reached America and destroyed chestnut trees in the East. Normally, because of its destructiveness, epidemic disease is sporadic and time an important factor. Endemic disease differs from epidemic disease in that the number of daughter infections per parent infection does not so far exceed 1 that, with infection constantly present, the level of disease would endanger the survival of host and pathogen. Host and pathogen coexist in balance.

This balance in endemic disease implies selection pressure in opposite directions. In one direction, selection pressure from disease increases resistance. In the opposite direction is a selection pressure which we identify with the disadvantage to the host of excess resistance; it is a selection pressure operating when there is no disease or not enough disease to matter. Without this opposing pressure, the host population would swing one-directionally to a state of high resistance and no disease. The effect of excess resistance results from the difference in metabolism and fitness between susceptible and resistant host plants; and the difference, we surmise, is greater when the disease is caused by fungi than when it is caused by less complicated viruses.

The state of no disease appearing in the title of this section and discussed in its second paragraph is in effect a state of immunity to disease in the population—population immunity for short. So too a state of low levels of endemic disease is population resistance. The words, immunity and resistance, properly apply to individual plants. Immunity is absolute. An individual cabbage plant is immune from the wheat stem rust fungus; it is a nonhost. Resistance is partial. Great resistance means small susceptibility. If the individual plants are immune, the population is also immune. But a population can be immune even if the individual plants are not. This is easily illustrated by reference to iR in Eq. (5.1). Immunity of individual plants means $R = 0$. Population immunity means $iR < 1$. Resistance of individual plants means that R is small. Population resistance means that iR does not greatly exceed 1. This can happen even if R is moderately large, provided that i is correspondingly very short (i.e., removals are quick). When, in Section 5.9 it was said that harmful viruses were not found in ecologically dominant perennial plants in their natural habitat, it was not implied that the plants were individually immune, but only that the population was immune (or tolerant).

Chapter 6

The Spread of Disease. Time and Distance as Dimensions

6.1 THE SPREAD OF DISEASE AND THE MIGRATION OF PATHOGENS

Disease has spread when diseased plants occur where they did not occur before, either in the immediate past or at any time previously. The spread of disease implies the migration of pathogens.

A pathogen may enter a new country and spread there once and for all time. *Endothia parasitica* was first observed in chestnut trees in the New York Zoological Park in 1904, having arrived, presumably, from the Old World. By 1908 it had spread abundantly north to Massachusetts, west to Pennsylvania, and south to New Jersey; and was already known in Delaware, Maryland, and Virginia. Thereafter, within a few years, it destroyed almost totally the native American chestnut trees in the eastern United States. *Cronartium ribicola* reached eastern North America late last century and western North America early this century, to cause blister rust in white pines. Both in the East and the West it migrated over immense areas. More recently, *Peronospora tabacina*, the cause of blue mold of tobacco, migrated from its point of escape in western Europe, eastward and southward into Asia and North Africa. Many other examples are known of immigrant pathogens that have spread widely and destructively after their arrival in a new country or continent.

Alternatively, the spread may be seasonal and recurrent. *Puccinia graminis tritici* migrates annually on wheat from Texas and the south to Canada. *Phytophthora infestans* survives the winter in diseased potato tubers; these serve as sources of infection from which the fungus spreads from field to field in spring and summer.

There is no evidence to suggest that once-and-for-all-time spread and

131

seasonal, recurrent spread are essentially different in any way; and we shall discuss them together.

6.2   SPREAD A FEATURE OF EPIDEMIC DISEASE

By definition, endemic disease excludes spread. Endemic disease is disease constantly present; and the spread of disease implies that the pathogen has moved to where it was not previously present.

Disease which was purely endemic would cover a constant area of countryside. But, as shown in the previous chapter, disease cannot be purely endemic except over a restricted geographical area like an island. Disease must be in an unsteady state. Around the area of endemic disease on a continent there must be a boundary area of disease advancing or receding in waves, like waves lapping the shore of a tideless lake. In this boundary area there will either be an epidemic, with disease spreading, or an antiepidemic, with disease receding. There can be no consistent tendency to spread. If there is a consistent tendency for disease to spread, the disease at the front is not endemic but epidemic.

No more will be written in this chapter on endemic disease. It is to be understood that in discussing the spread of disease we are discussing epidemic disease.

In discussing the spread of disease two dimensions need particular attention: the dimension of time, because the disease is epidemic, and the dimension of distance, because that is what spread is about. Both Chapters 4 and 6 are about time as a dimension, i.e., both are concerned with rates. Chapter 6 differs from Chapter 4 in that it is concerned with distance as well.

6.3   THE RATE OF MULTIPLICATION (THE INFECTION RATE) AND THE RATE OF SPREAD OF DISEASE. DISPERSAL OF PATHOGENS

Chapter 4 dealt with the infection rate, which is the rate of multipli- cation or increase of disease, without reference to the area covered by the disease. The rate of spread depends on factors almost all of which are identical with those governing the infection rate.

If the host plants are very susceptible, disease will multiply very fast. It will also spread very fast. The more susceptible the host plants, the faster the multiplication of disease and the faster the spread of disease.

This holds for other factors too. Factors that favor multiplication of disease—a wet climate, if wetness favors infection, a warm climate, if warmth favors disease, and so on—are also factors that usually favor the spread of disease. Conversely, factors that restrict the multiplication of disease—more resistant host plants, a climate unfavorable to disease, and so on—are also factors that usually restrict the spread of disease.

Distance enters the multiplication and spread of disease in two ways only. The pathogen travels inside the plant and it travels outside the plant.

Travel inside a plant has only a small direct part in the geographical spread of disease, and we shall ignore it for the time being. It has indeed a major indirect part through its effect on the period of latency p and, via sporulation, on the infection rate R. These are matters that come into Section 6.8.

Travel outside the plant is dispersal; and dispersal is the direct way in which distance enters geographical spread. A study of dispersal distinguishes this chapter from Chapter 4.

6.4   SPREAD AND DISPERSAL. EFFECTIVE DISPERSAL

As we define spread, a pathogen spreads where it goes and infects. Spread is essentially migration. Thus when we say that *Peronospora tabacina* spread within a few years from its point of departure in western Europe to Asia and North Africa, we mean that it migrated within a few years from western Europe to Asia and North Africa. It is not implied that a few spores escaped and, in a single act of dispersal, crossed Europe and the Mediterranean. It is implied only that conditions favored disease—there were abundant susceptible host plants, and the climate permitted infection—and the pathogen made its way by means of an unrecorded number of acts of dispersal until it reached Asia and North Africa.

Dispersal, with reference to spores, is the process from takeoff to deposition. More generally, and not just for spores in particular, dispersal is the movement of units from a lesion to healthy, susceptible tissue, if the pathogen causes local lesions, or from infected plants to healthy, susceptible plants, if the pathogen infects systemically. Units of dispersal have already been defined in Chapter 1. They include fungus spores, bacterial cells, virus particles, clumps of fungus spores held and dispersed together in mucilage, infected leaves or pieces of leaves blown about by wind or moved by any other agency, virus doses transmitted as entities by vectors, and so. Nothing is excluded *a priori*. Man is included as an efficient vector; and the carriage by man of an infected plant, or part of a plant, from one continent to another, to infect a healthy plant there, is an act of dispersal.

For the purpose of this chapter dispersal is limited to effective dispersal. A spore from a leaf falling to the ground and perishing there would not in this chapter be considered to have been dispersed; nor would a spore carried to an immune host, like a wheat stem rust spore carried to a cabbage leaf, be considered to have been dispersed. Dispersal in the limited sense in which we shall use the word here implies subsequent infection.

6.5 FOCI OF DISEASE

Disease does not occur uniformly distributed in space. There is more disease in some areas than others; that is, disease occurs in foci.

Foci are areas of higher than average disease. To give in full the definition adopted by the British Mycological Society (Anonymous, 1953), a focus is the site of local concentrations of diseased plants or disease lesions, whether about a primary source of infection or coinciding with an area originally favorable to establishment, and tending to influence the pattern of further disease.

Among the earliest observers of foci were Heald and Studhalter (1914). While *Endothia parasitica* was destroying indigenous chestnut trees of the eastern United States, the epidemic moved along a broad front. Often miles ahead of this front were "spot" infections—new foci —which allowed the pathogen to advance with long leaps, and so accelerate the infection of its host.

Foci of potato blight caused by *Phytophthora infestans* have often been studied. Brenchley and Dadd (1962) took infrared photographs from the air. These show, first, that during the early part of an epidemic *P. infestans* occurred in well-defined foci which stand out clearly against the general background of the rest of the field; and, second, that daughter foci were not markedly clustered near the initial focus. The daughter foci were scattered through the crops. These two features are important for the understanding of the dispersal process.

Later in an epidemic, when the general level of disease is high, foci becomes inconspicuous and small, as overcrowding inevitably forces them to overlap and hem one another in.

There is a large literature of foci. Some of it was reviewed by Van der Plank (1963, pp. 77–89), who also gave a diagram of foci of cacao swollen shoot disease based on information by Thresh (1958).

6.6 VARYING GRADIENTS OF DISPERSAL. MIGRATION AND COLONIZATION

The two features of potato blight foci seen by infrared photography from the air—the foci are clearly defined, and daughter foci are not markedly clustered near the parent focus—show that *Phytophthora infestans* disperses along at least two different gradients. It disperses frequently over a very steep gradient and less frequently over a shallow gradient. Dispersal over a shallow gradient starts new foci; dispersal over a steep gradient enlarges them.

Figure 6.1 illustrates five experimental gradients of infection by *P. infestans* coming from a point source of inoculum. The gradients are very steep; at low levels, disease varies inversely as approximately the fifth power of distance from the source of inoculum. Two theoretical curves, marked *G*, have been added to represent Gregory's (1945) equation for dispersal in air under conditions of low atmospheric turbulence. For convenience we have kept to the parameters that Gregory himself used.

Dispersal along a very steep gradient can enlarge existing foci, keeping the advancing boundary of the focus clear and well defined against the background of the rest of the field. The steeper the gradient, the clearer is the definition. But dispersal along a steep gradient is unlikely to start daughter foci widely separated from their parents; the steeper the gradient, the less the likelihood of daughter foci widely separated from their parents.

Phytophthora infestans can also disperse along shallow gradients to start scattered foci widely separated from the source of inoculum. Schrödter (1960) summarized the evidence in the literature about this, and quoted observations by various workers which suggest that sporangia of *P. infestans* have a flight range of more than 60 km.

Van der Plank (1967b) analyzed information on a photograph of the first potato blight outbreak photographed by Brenchley (1964): an outbreak near Lakenheath, west Suffolk. A blighted dump of potatoes in a ditch separating two fields of potatoes started the outbreak. From this focus, four new foci were established, as daughter and granddaughter foci, at different dates before July 27. They were 30, 150, 25, and

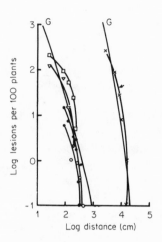

Fig. 6.1 The amount of potato blight caused by *Phytophthora infestans* at varying distances from the source of inoculum. From left to right, the first two experimental curves are drawn from data of Waggoner (1952), the next two from data of Limasset (1939), and the last from data of Bonde and Schultz (1943). The two curves marked G are calculated from Gregory's (1945) equation for dispersal in conditions of low turbulence. The data of Limasset and Waggoner were given as percentage of disease, and have been transformed into the estimated number of lesions per 100 plants by means of Eq. (1.1). (Modified from Van der Plank, 1960, p. 255.)

40 yards from their respective sources, all in an apparently uniform field. There were no other sources of blight in the neighborhood; in the whole survey area of 70 miles^2 only seven outbreak centers were found at that early date in the summer. The problem is this: There were four new foci, 30, 150, 25, and 40 yards from, and no nearer, the source of infection. What is the probability of this occurring with dispersal following Gregory's equation and exemplified by the *G* lines in Fig. 6.1? The answer is, less than 10^{-8}. (In this calculation it is assumed, probably realistically enough, that the field was uniform, the parent foci circular and 2 m across, and the spores came from the center of the foci.)

Evidently, steep gradients that give sharp boundaries to foci are incompatible with the establishment of daughter foci all far from these boundaries. Shallow gradients are needed for this.

What is the relative frequency of dispersal along steep and along shallow gradients? If we assume that each daughter focus starts from a single lesion, the average number of lesions in a focus is also the average ratio of the number of acts of dispersal which enlarge existing foci to the number which start new foci, i.e., it is the ratio of the number of acts of dispersal along a steep gradient to the number along a shallow gradient. There are often 1000 blight lesions on a single, heavily infected potato plant, and many infected plants in a single focus. The ratio is therefore great, and the relative frequency of dispersal along shallow gradients small.

Long-distance dispersal along shallow gradients, although relatively rare, is nevertheless important for the survival of the pathogen. Either steep gradients only or shallow gradients only would serve the pathogen badly. Steep gradients only would confine the pathogen to an overcrowded focus in which further expansion was possible only along a boundary of ever-diminishing extent in proportion to the area of the focus. Shallow gradients only would carry sporangia far and wide, so that many would fall where there were no host plants. A mixture of shallow and steep gradients means that the pathogen dispersing along steep gradients could colonize any susceptible plants or fields it found after dispersing along shallow gradients.

Foci of plant disease are simply examples of colonization following migration, a pattern that seems universal in nature. Migration along shallow gradients over relatively long distances is followed by colonization along steep gradients over relatively short distances. Winged aphids find susceptible host plants; generations of wingless aphids colonize the find. Man himself in his migrations conformed with this pattern.

To bring out the point that the focal development of disease is probably nothing more than a nature-wide pattern of population increase, we have oversimplified matters by dealing with only two sorts of gradient, steep and shallow. Almost certainly, there are many sorts of gradients, perhaps an infinite number of them, all meeting a particular need for the efficient increase of the pathogen. Attempts to find a characteristic form of dispersal in a pathogen—attempts to find that *Phytophthora infestans* disperses in this way, and *Puccinia graminis* in that way—are almost certainly doomed to fail.

Sporangia of *P. infestans* are carried in water. They are carried in air. These statements are themselves oversimplifications. As regards dispersal in water, sporangia may be carried by dew on a still day. They may be carried in driving rain in a gale. They may germinate and become motile as zoospores. As regards dispersal by air, Hirst (1958) trapped sporangia in the morning, noon, and afternoon, on rainy days and dry. To choose just one factor, the turbulence of the air must have varied greatly, and the gradient with it. And water and air are not the only agents. Man is an efficient vector, carrying infected seed potatoes far and wide. One of Brenchley and Dadd's (1962) photographs shows blight strongly developed along the paths of wheels of spray machinery . . . .

Consider two other pathogens. Heald and Studhalter (1914) and Leach (1940) believed that *Endothia parasitica* spread among chestnut trees in at least two ways. Migratory birds carried sticky pycnospores on their beaks to start distant foci. Airborne ascospores which the fungus produces abundantly enlarged the focus.

Hemileia vastatrix causes leaf rust in coffee. Nutman *et al.* (1960) showed that uredospores are dispersed by rain splash, but only over short distances. Yet there has long been evidence that the fungus does in fact disperse far. It has no other host than coffee and is not seed borne, but it soon appears in plantations grown from seed far from any other coffee. Crowe (1963) suggested on circumstantial evidence that more distant dispersal is by insects. The larvae of two species of midge eat spores of *H. vastatrix* in the rust pustules. These larvae are in turn parasitized by two species of Hymenoptera which Crowe regards as the vectors of the fungus. When rust is abundant, so are the midge larvae and their winged parasites. Crowe trapped parasites away from infected coffee and found spores of *H. vastatrix* on them. More recently, *H. vastatrix* has been found in coffee plantations in Brazil, thus being known in the New World for the first time. *Hemileia vastatrix* jumped the ocean—whether it was carried by man or by wind is irrelevant to our narrative—and has begun to colonize Brazil. Coffee leaf rust illustrates two matters central to the study of spread. First, a study, however detailed, of what is presumably the common method of dispersal, by rain splash, gives a distorted picture of the fungus' ability to spread. Second, one dispersal—or at most a few dispersals—over a long distance to Brazil has given the fungus the opportunity to colonize an enormous area of coffee leaves; the fungus can follow a rare long-distance dispersal along a shallow gradient by billions of dispersals along steep or relatively steep gradients. The frequency of dispersal, as measured, e.g., by spore trapping, is a poor guide to the integrated pattern of spread.

6.7   THE MULTIPLICATION AND SPREAD OF *Phytophthora infestans* SEEN AS CONCURRENT PROCESSES

Figures 6.2 and 6.3 show two sides of the same picture of infection by *Phytophthora infestans*. The data (Anonymous, 1954) are for infec-

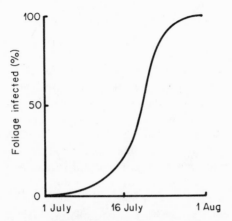

Fig. 6.2   The progress of blight caused by *Phytophthora infestans* in foliage of the potato variety Bintje in the sand areas of the Netherlands in 1953 (Anonymous, 1954). The percentages refer to the total foliage of all fields of the variety, in the stated areas. Blight was first recorded at the end of May, but the percentage during June was too small to be recorded by the graph.

tion in 1953 of the very susceptible potato variety Bintje in the sand areas of the Netherlands.

Figure 6.2 shows the progress of the blight epidemic in terms of the percentage of foliage infected. The S-shaped disease progress curve

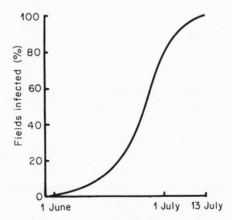

Fig. 6.3   The spread of *P. infestans* from field to field of the potato variety Bintje in the sand areas of the Netherlands in 1953 (Anonymous, 1954). The spread is measured by the increase in the percentage of fields in which *P. infestans* was found.

shows the course of destruction of the total foliage of the variety Bintje in the area.

Figure 6.3 shows the progress of the blight epidemic in terms of the percentage of fields of the variety Bintje infected. The curve shows how the fungus spread from field to field. It measures spread, with fields as units of distance, whereas Fig. 6.2 tells us nothing about distance in spread.

Figure 6.2 is concerned with the basic topic of Chapter 4: how an epidemic progresses with time. The data given are the material from which infection rates can be calculated. All spread of the fungus, be it from field to field, or from plant to plant in the same field, or from leaf to leaf on the same plant, is relevant.

Figure 6.3 is concerned with a basic topic of Chapter 6: how an epidemic spreads with time. Only the spread of the fungus from field to field is shown, because this is how distance was recorded in the epidemic.

The processes of infection and spread occurred concurrently. Blight was first noticed, and the first field recorded as infected, at the end of May. (The level of disease during June was too low to be recorded graphically in Fig. 6.2.) By the middle of July, *P. infestans* was recorded in all fields, although in most of them the percentage of foliage infected was still small. During the second half of July the destruction of the foliage was completed and the epidemic in the variety Bintje was at an end by August 1.

6.8   A CORRELATION BETWEEN THE INFECTION RATE AND THE RATE OF SPREAD OF DISEASE

What is the quantitative relation between the infection rate r and the rate of spread of disease? What is the connection between r and the rate at which the front of an epidemic advances? Two preliminary matters need to be discussed.

Consider the advance of an established epidemic. After it was well established in the West, *Cronartium ribicola* advanced 20 miles a year in white pines (see Section 5.1). *Edothia parasitica* advanced at about the same rate in native chestnuts in the East. In Georgia in 1970 *Helminthosporium maydis* advanced 20 miles a day in maize. Because pathogens disperse along gradients, most inoculum for the advance comes from near the front, disease in the rear being mainly removed, removal having the meaning given in Section 4.2. Inoculum comes from a source which in size is largely independent of time. Infection ahead of the front is therefore largely at "simple interest," the rate being proportional to R.

The rate of spread of disease can be brought into the discussion thus: Suppose that it requires X acts of dispersal for the pathogen to advance one mile. Then if X is independent of R, an assumption that is a good approximation when the epidemic is fast and iR large, the rate of spread, which is one mile in the time needed for X acts of dispersal, is proportional to R. If R is halved the frequency of acts of dispersal is halved, and it will take twice as long for X acts of dispersal to occur. If R is halved it will take twice as long for the pathogen to spread 1 mile; The rate of spread over distance will be halved. Put generally, the rate of spread is proportional to R.* Equation (4.3) relates r and R; therefore it relates the multiplication rate (i.e., the infection rate) to the rate of spread.

Equation (4.3) shows that large changes of R can go with small changes of r. R is highly geared to r. The gearing of R to r increases as pr increases. A high value of r or a long period of latency p increases the change in rate of spread for a given change in rate of multiplication of disease.

The gradient of disease in a well-established epidemic depends on the infection rate. The gearing implicit in Eq.(4.3) makes this so. *From any given level of disease* in an established epidemic the gradient will be flatter as the infection rate is faster, other things being equal. The speed of an epidemic determines its compactness at any given level of disease at the center. Slow epidemics can be expected to be less diffuse than fast epidemics of the same disease, even after there has been time

---

*In Chapter 5, iR was the average number of daughter lesions per parent lesion, i.e., the number produced by the parent lesion during its whole lifetime or, what comes to the same thing, its whole period of infectiousness. For Chapter 6, iR is the average number of effective dispersals from a lesion during its whole period of infectiousness. The meanings in the two Chapters are identical. But we are concerned in Section 6.8 with rates, not numbers, and therefore with R, not iR. All the meanings trace back to Eq. (4.2). It is the ability of this differential difference equation to describe ordinary processes of infection and to be used consistently for the different topics of Chapters 4–6 that makes our system of epidemiology more flexible for our purposes than the system used in medical epidemiology that requires the estimation of removal rates. For systems of equations involving removal rates see, e.g., Bailey (1957).

enough to compensate for slower progress and allow for a comparable level of disease to occur. The inherent error in the assumption that X is independent of R is unimportant in this comparison, to the extent that the error can only increase the difference between fast and slow epidemics.

This tentative and uncomplicated treatment of the rate of spread requires that infection ahead of the front (though not necessarily at or behind the front) should be at simple interest, which in turn requires that the epidemic should be well established with removal of disease behind the front as a substantial source of infection ahead. Young epidemics, just starting, develop more nearly at compound interest; and the mathematical difficulties of relating the infection rate and the rate of spread have remained insuperable.

Chapter 7

Genetics of Host–Pathogen Relations

7.1 INTRODUCTION

In previous chapters plant disease was seen primarily in terms of populations of the pathogen. Here we continue the emphasis on populations, and study host–pathogen relations mainly in terms of the population genetics of the pathogen. That is, we study the frequency in the population of genes for virulence or other forms of pathogenicity, and try to relate this frequency to the genetic constitution of the host plants.

The current definition of immunity and resistance is adopted. Immunity is complete and absolute. Resistance is partial and relative. Immunity is the common condition. Any one given plant is immune from most of the known pathogens; i.e., it cannot act as host to them. Against the relatively few pathogens that can attack it, the host plant shows varying degrees of resistance.

Although immunity is the more usual condition, it has been much less studied than resistance. Most of the available data in publications are about resistance. This is reflected here. Resistance is discussed first, and immunity thereafter. This procedure, dictated by convenience, may in fact be logical. There is evidence that immunity differs essentially only in degree, and what will be said about immunity links closely with what will be said earlier about resistance.

7.2 THE FUNDAMENTAL CLASSIFICATION OF RESISTANCE. DIFFERENTIAL AND UNIFORM INTERACTIONS BETWEEN HOST AND PATHOGEN. VERTICAL AND HORIZONTAL RESISTANCE

Races of the pathogen may interact differentially with varieties of the host. For example, race 1 of *Phytophthora infestans* attacks potatoes with the gene R_1 but not those with the gene R_2, whereas race 2 attacks those with the gene R_2 but not those with the gene R_1.

It is equally possible for races of the pathogen to interact uniformly (i.e., nondifferentially) with varieties of the host. Thus when *P. infestans* attacks potato varieties without *R* genes, isolates of the fungus differ significantly in the rate at which they can attack, and varieties of potatoes differ significantly in the rate at which they are attacked. But, on evidence presented in Section 7.13, there is no significant differential interaction between the fungus isolates and the potato varieties.

When differential interactions are present, we shall call the resistance vertical; when they are absent we shall call the resistance horizontal. The merits and demerits of various terms will be discussed later. For the moment we are concerned with the primary classification of resistance into differential and nondifferential resistance.

Gene-for-gene hypotheses, such as Flor's, are conceivable only if resistance is differential. One could define vertical resistance as resistance for which a gene-for-gene hypothesis is conceivable, and horizontal resistance as resistance for which a gene-for-gene hypothesis is inconceivable.

A differential interaction between host varieties and pathogenic races allows the host to exert direct selection pressure on races of the pathogen; there is directional selection of pathogenic races in favor of those compatible with the host. In the absence of a differential interaction between host and pathogen, any selective advantage or disadvantage a pathogenic race may have is necessarily only indirectly and remotely related to the variety of the host. A differential interaction between host varieties and pathogenic races is the source of the boom and bust cycles for which vertical resistance is notorious. Conversely, the absence of a differential interaction is the source of the stability that one associates with horizontal resistance.

The division between differential and nondifferential interactions is clear and unambiguous. There is a straight either–or situation: there is either a differential interaction between host varieties and pathogenic races or there is not. Judged by host–pathogen interaction, resistance is naturally of only two kinds, without intermediates. Obviously there can exist in any single host a mixture of the two kinds. Indeed, as we shall see, it seems certain that vertical resistance never occurs alone; that is, it is always mixed with horizontal resistance. But the occurrence of mixtures is no reason for blurring the distinction between the kinds of resistance, any more than coexistance in the earth's crust is a reason for blurring the distinction between iron and aluminum.

The extent of current confusion is evident from a recent review. Wolfe (1972) writes that "the complexity of the interaction between the host and the pathogen suggests that there may be no fundamental difference between vertical and many types of horizontal resistance. These, and similarly defined terms, may be descriptions of the points of a continuum of immunity and susceptibility in the host, matched by a corresponding continuum between nonvirulence and virulence in the pathogen." Wolfe allows complexity to blur distinctions. There can be a continuum of mixtures of vertical and horizontal resistance, just as there can be a continuum of mixtures of iron and aluminum in the soil. There can be a continuum of efficacy of vertical resistance or of horizontal resistance. But the complexities and continua relate to matters other than the distinction between differential and uniform interactions.

Almost everything discussed in this chapter is cleft between matters arising from a differential interaction between varieties of the host and races of the pathogen and those arising from a uniform interaction. We begin with differential interactions, i.e., with vertical resistance.

7.3 FLOR'S GENE-FOR-GENE HYPOTHESIS

In simple language a differential interaction between varieties of the host and races of the pathogen means that a variety of the host has its "own" races of the pathogen: races that are differentially adapted to it, though not necessarily exclusively adapted to it. Correspondingly, a race of the pathogen has its "own" varieties of the host, to which it is differentially, though not necessarily exclusively, adapted. This opens the possibility for genes in the pathogen to correspond with genes in the host; and a correspondence in a one-for-one relation is Flor's gene-for-gene hypothesis.

Flor's enunciation of his gene-for-gene hypothesis was one of the really important events in the history of plant pathology. A modern review by Flor (1971) may be consulted for recent developments and general references.

Flor proposed his hypothesis for the relation between resistance in flax (*Linum usitatissimum*) and virulence in the rust fungus (*Melamp-*

sora lini) that attacks it. Since then the hypothesis has been extended by others to the host–pathogen combinations listed in Table 7.1. Of these combinations, only the flax rust, potato blight (Toxopeus, 1956; Person, 1959), and wheat stem rust (Loegering and Powers, 1962; Green, 1966; Williams *et al.*, 1966; and Kao and Knott, 1969) systems will be discussed here.

In his work on flax rust, Flor was the first to study the genetics of both members of a host–pathogen system. From his results he concluded that for each gene determining resistance in flax there is a specific and related gene determining pathogenicity in the rust fungus. In flax varieties possessing one gene for resistance to the avirulent parent race, pathogenicity was conditioned by one gene in the fungus. In the flax varieties possessing two, three, or four genes for resistance to the avirulent parent race, pathogenicity was conditioned by two, three, or four genes in the fungus. The hypothesis, that for each resistance gene in the host there is a matching or reciprocal gene for pathogenicity in the fungus, is the simplest that fits these facts.

Resistance genes occur as multiple alleles in five loci. The symbols *K, L, M, N,* and *P* designate genes in the five loci. Of the known resistance genes, one lies in *K,* twelve in *L,* six in *M,* three in *N,* and four in

TABLE 7.1

Host–Parasite Systems for Which Gene-for-Gene Identity Relationships
Have Been Suggested

Avena	–	*Helminthosporium victoriae*
Avena	–	*Puccinia graminis avenae*
Avena	–	*Ustilago avenae*
Coffea	–	*Hemileia vastatrix*
Helianthus	–	*Puccinia helianthi*
Linum	–	*Melampsora lini*
Lycopersicon	–	*Cladosporium fulvum*
Malus	–	*Venturia inaequalis*
Solanum	–	*Phytophthora infestans*
Solanum	–	*Synchytrium endobioticum*
Triticum	–	*Erysiphe graminis tritici*
Triticum	–	*Puccinia graminis tritici*
Triticum	–	*Puccinia recondita*
Triticum	–	*Puccinia striiformis*
Triticum	–	*Tilletia caries*
Triticum	–	*Tilletia contraversa*
Triticum	–	*Ustilago tritici*

the P locus. Resistance genes in each locus are identified by numerical superscripts, e.g., L^4.

The genes for pathogenicity are recessive, and are identified to correspond with the resistance gene. Thus, aL^4 is the gene for pathogenicity matching the resistance gene L^4.

Flor's hypothesis is purely a hypothesis of identities. He identifies the resistance gene, and he identifies the matching gene for pathogenicity. Identities have their limits. We identify a man by his name and address. Joseph J. Smith lives at 5 Waupelani Drive. That will identify him in a telephone directory. If needed, we can also identify the town and country. But the name and address tell us nothing about his qualities. Is he clever or stupid? Is he worth hiring for a job? That we are not told. So too we identify a resistance gene by its number and its locus. If needed, we can also identify the chromosome and host plant. But the number and locus tell us nothing about the gene's qualities. Is it, e.g., likely to be useful to the plant breeder or not?

7.4 A Second Gene-for-Gene Hypothesis. Gene Identity and Quality

There is no known limit to the number of gene-for-gene relations needed to describe the interactions between host and parasite. Flor's hypothesis deals with gene identity. A second gene-for-gene hypothesis is introduced here; it deals with gene quality. There may be other relations still to be discovered.

There are ten or more genes for resistance to potato blight : R_1, R_2, . . ., R_{10}. We can identify them, and the first four R_1, R_2, R_3, and R_4 have been well studied. Are they of equal value to a potato breeder? Suppose that in the 1950's, when enthusiasm for R-gene resistance was at its peak, a potato breeder had put the available numbers in a hat and drawn at random, to see what gene he should incorporate in his new potato varieties. Would he have been equally lucky with any gene he drew? The answer is, definitely not. If he had drawn R_4 he would have drawn a gene nearly useless to him. If he had drawn R_1 he would have picked a gene which, in the 1950's at any rate, was temporarily very valuable, even when used on its own.

So too there are some 25 Sr genes in wheat for resistance to stem rust. To see the matter in the wisdom of retrospect, suppose that a quarter of a century ago a breeder of spring wheat in North America had put the numbers in a hat and chosen randomly to decide what gene to use. Would it have made no difference to him what number he chose? Could he, e.g., have picked Sr_5 as usefully as Sr_6? The answer is, definitely not. In retrospect it is clear that, added to Sr_{9d} which was then prevalent in spring wheat, Sr_6 was far more useful than Sr_5 would have been.

Granted these statements, which we shall substantiate in the next few sections, it follows that resistance genes have qualities as well as identities. Flor's hypothesis, we have noticed, deals with identity. We shall deal with quality, the specific quality being the amount of protection against disease the resistance gene can give. This quality, our hypothesis states, is also on a gene-for-gene basis.

Let us formulate the second gene-for-gene hypothesis. In host–parasite systems in which there is a gene-for-gene relationship, the quality of a resistance gene in the host determines the fitness of the matching virulence gene in the parasite to survive when this virulence is unnecessary; and reciprocally the fitness of the virulence gene to survive when it is unnecessary determines the quality of the matching resistance gene as judged by the protection it can give to the host.

Flor introduced his hypothesis on the limited basis of the flax rust system, and left it to be extended later to other host–pathogen systems. So too we shall limit discussion primarily to five resistance genes: R_1 and R_4 in the potato, and Sr_5, Sr_6, and Sr_{9d} in wheat. If we can demonstrate a quality difference between any two R genes or any two Sr genes, and also demonstrate that this quality difference is related to a quality difference in the pathogen on a gene-for-gene system, we shall have established the case for a second gene-for-gene hypothesis.

7.5 POTATO BLIGHT: THE RESISTANCE GENES R_1 AND R_4, AND THE FREQUENCY OF VIRULENCE ON THEM IN POPULATIONS OF *Phytophthora infestans*

The gene R_4 has not been successfully used by potato breeders on its own, although it has been used occasionally in combination with other R genes. The gene R_1 has often been used, and, at least for a time, it successfully protected potato varieties against blight. The difference between these two R genes is explained by the fact that virulence on R_4 preexisted abundantly in populations of *Phytophthora infestans*, whereas virulence on R_1 did not. Consider the evidence.

Hogen-Esch and Zingstra (1969) compiled an international list of potato varieties, giving information *inter alia* about R genes. The information is not always complete. But in order to avoid any possible subjective bias, the information has been used as it stands, without addition or amendment. It need only be commented that any additions or amendments the author could make would not affect the analysis that follows.

None of the varieties in the list has the gene R_4 alone; that is, no breeder has successfully used this gene on its own. However, it has been used in five varieties in combination with other R genes. (It was used three times in combination with R_1, once with R_2, and once with R_3.) In contrast with this, the gene R_1 has been used on its own in 54 of the listed varieties and in combination with other R genes (mostly R_3) in 19 varieties. For the purpose of a statistical test we need use only the ratios $0:5$ and $54:19$. The disparity between them is highly significant (with P less than 0.001). The gene R_4 needed the help of other genes more than R_1 did. By comparing ratios the possibility that the failure of R_4 was due to lesser availability or to adverse linkages is automatically excluded. There is also independent evidence about this.

The gene R_4 has been known from the 1930's and almost as long as R_1. It first appeared commercially in the potato variety Virginia. Virginia has the genes R_1 and R_4, and was released in 1950. The variety Vertifolia, with R_3 and R_4, followed soon after. It normally takes at least 10 years for a breeder to select and launch a new variety, and we can safely assume that potato breeders were using the gene R_4 in the 1930's. Nor was its use restricted to a few breeders. The five listed varieties with R_4 in combination with other R genes came from countries as widely separated as Japan, Denmark, Estonia, Poland, and Mexico. The use of R_4 in combination, though not alone, shows that it was not linked with adverse agronomic qualities. Indeed, in the general descriptions the varieties with R_4 were often highly praised for agronomic performance.

Consider now the frequency of virulence in populations of *Phytophthora infestans*. For the present analysis the information needed is about the populations before they were changed by the introduction of potato varieties with R genes, i.e., about wild-type populations.

Frandsen (1956) in the early 1950's collected 34 isolates of *P. infestans* from farmers' potato fields in northwestern Germany. The potatoes had no R genes. Yet all the isolates were virulent on R_4. Virulence preexisted. In the environment sampled by Frandsen no potato breeder could have used R_4 to protect potato varieties from blight, because it would have made no difference to the variety's blight resistance whether it contained R_4 or not. This illustrates one of the points in our gene-for-gene hypothesis. The great fitness of virulence on R_4 to survive even when it is unnecessary shows that the quality of R_4, judged by the protection it can give, is poor.

Less extreme were the findings of Graham (1955) who surveyed populations of *P. infestans* in Canada at about the same time as Frandsen worked in Germany. Potato fields in Canada contain no R_4; this gene has not been used commercially there. Yet of 68 isolates from farmers' fields 39, i.e., more than half, were virulent on R_4. Virulence preexisted abundantly—enough to make the gene R_4 practically worthless to potato breeders.

This pattern of abundant virulence on R_4 is worldwide. Table 7.2, based largely on a table by Van der Plank (1968), shows this. In most of the countries mentioned R_4 does not occur in potato fields because breeders have not successfully used it. The populations of *P. infestans* with their abundant virulence on R_4 do not reflect an adaptation to this resistance gene; they testify to the fitness of the virulence to survive even when it is unnecessary.

Consider now the gene R_1. Cultivars with this gene have become widely grown, and populations of *P. infestans* have become adapted to it. We are not at present concerned with the process of adaptation— the process that is the downfall of vertical resistance—but with the

TABLE 7.2

Percent Frequency of Isolates of *Phytophthora infestans* Virulent on Resistance Gene R_4 in Various Countries

Country	Percent frequency
Canada	57
Columbia	small
England	86
Germany	95
Mexico	38
Netherlands	72
Northern Ireland	96
Rhodesia	94
Scotland	90
Sweden	large
United States	63

ancestral, wild-type populations before the process began. The first successful cultivars with R_1 were released in the 1940's—Kennebec, the most successful of R_1 types in North America, was released in 1948—but were not common until after 1950.

The infected fields from which Frandsen (1956) isolated *P. infestans* in the early 1950's were free from R_1 as well as R_4. Of his 34 isolates one was virulent on R_1. Of the isolates that Graham (1955) studied in Canada 56 came from potato fields without R_1 and two from tomato fields. None was virulent on R_1.

The disparity between the ratios 1 virulent : 33 avirulent on R_1 and 34 virulent : 0 avirulent on R_4 in Frandsen's data is highly significant statistically, as is the disparity between the corresponding ratios 0:58 and 39:29 in Graham's data.

The data of Frandsen and Graham are the only available sets that can be accepted as illustrating virulence on R_1 in wild-type populations. Other early surveys were made, but the publications did not specify the genotype of the host plants from which isolates were taken. (It is obvious from the context of these surveys that isolates were made from lesions on potatoes with R_1). Later surveys often reflected the growing popularity of R_1 types of potato, and the adaptation of *P. infestans* to them. Nevertheless, the general evidence is wholly consistent with the findings of Frandsen and Graham that virulence on R_1 preexisted rarely. Toxopeus (1956) in the Netherlands found virulence on R_1 to be so rare that he suggested it arose by mutation from avirulence during the course of the season. As to other countries, preexisting virulence was automatically sampled by the first resistant potato cultivars to be released. Cultivars with R_1 were everywhere resistant when they were first released. Even to this day virulence on R_1 tends to be rare where this resistance gene has not been used. In the Netherlands R_1 is common in cultivars in the north but rare in the south. This is reflected in the frequency of virulence on R_1. It is common in populations of *P. infestans* in the north but rare in the south (Mooi, 1968, 1969; and in private communication, 1972). Typically, virulence on R_4 is common both north and south even though it is unnecessary.

In relation to the second gene-for-gene hypothesis it has been our concern to prove that there is a difference in quality between the resistance genes R_1 and R_4, as shown by the greater use potato breeders have made of R_1 and, by implication, the greater protection against blight it has given; and that there is a difference in quality between the virulence on R_1 and R_4, as shown by the greater fitness of virulence on

R_4 to survive when it is unnecessary. The remaining points, that the virulences on R_1 and R_4 are what they are because the resistance genes R_1 and R_4 are what they are and that the virulence genes themselves affect the fitness, will be taken up at the end of the next section, in relation to the corresponding points with the Sr genes.

7.6   WHEAT STEM RUST : THE RESISTANCE GENES Sr_5, Sr_6, AND Sr_{9d}, AND THE FREQUENCY OF VIRULENCE ON THEM IN POPULATIONS OF *Puccinia graminis tritici* IN CANADA

The gene Sr_5 has occasionally been used on its own in wheat cultivars not noted for great stem rust resistance. But mostly it has been used in combination with other Sr genes. (It occurs in the North American wheats Thatcher, Canthatch, Kanred, Pembina, Reliance, and Manitou, and in the Australian Timgalen.) On its own it is nearly ineffective (Green, 1971). It thus has much in common with the gene R_4 in potatoes. The genes Sr_6 and Sr_{9d} on the other hand have had distinguished careers. The gene Sr_{9d} saved North American spring wheats from race 56 of stem rust, from the 1930's through the 1940's; and the gene Sr_6, added to Sr_{9d}, has saved wheat since the early 1950's.

The most complete information comes from Canada, where Green (1971) summarized data for the years 1950 through 1969. The earlier data were for conventional standard races. The later data were obtained from Green's much more rewarding virulence formula for designating races. Races are classified according to the effective/ineffective host genes. Thus the common Canadian race $C18$ has the virulence formula 6, 8, 9*a*, 9*b*, 13, 15/5, 7, 9*d*, 10, 11, 14, 16. That is, race $C18$ is avirulent on genes Sr_6, Sr_8, Sr_{9a}, Sr_{9b}, Sr_{13}, and Sr_{15}, but virulent on genes Sr_5, Sr_7, Sr_{9d}, Sr_{10}, Sr_{11}, Sr_{14}, and Sr_{16}.

In the early 1950's, with a peak in 1952 and 1953, there was a severe epidemic of stem rust in Canadian wheat. The stem rust race mainly concerned was 15*B*-1 (old style). Then the new wheat cultivar Selkirk was introduced. It had the gene Sr_6 in addition to the old gene Sr_{9d} which had protected Canadian wheat in the 1940's but was ineffective against race 15*B*-1. This addition of gene Sr_6 proved effective, and Selkirk kept stem rust in check. At the start the omens were not auspicious. In 1953 a race appeared—race 15*B*-3—which could attack Selkirk. But it and other virulent races never became abundant, and the resistance given by genes Sr_6 and Sr_{9d} was maintained.

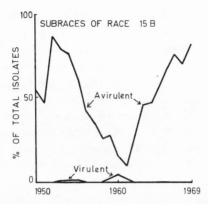

Fig. 7.1   The number of isolates of subraces of race 15*B*-1 in Canada, expressed as a percentage of isolates of all races. The upper line is for subraces avirulent on the gene Sr_6, and the lower line for virulent subraces. Except in 1960 virulence on Sr_6 was rare within race 15*B*. Data of Green (1971).

Figure 7.1 analyzes the data of Green (1971) for the six known Canadian subraces of race 15*B*.* Four of these subraces (15*B*-1, 15*B*-1*L*, 15*B*-1*XL*, and 15*B*-4) are avirulent on the gene Sr_6, and two (15*B*-3 and 15*B*-5) are virulent. All six subraces are virulent on genes Sr_5 and Sr_{9d}, so the distinction is essentially between avirulence and virulence on gene Sr_6. Figure 7.1 shows that the unexpected happened. The virulent subraces that had the selective advantage of being able to attack Selkirk never became abundant. After a minor flareup in 1960, when six virulent isolates were obtained in the Prairie Provinces, they were unrecorded in the following nine years of the survey. The avirulent subraces on the other hand, despite their selective disadvantage of not being able to attack Selkirk, were the subraces that flourished. (That is, they flourished in the susceptible wheat varieties used in the Canadian wheat stem rust survey to indicate what sort of inoculum was about.) In other words, the gene Sr_6, with Sr_{9d} as background, remained effective. Ordinarily, it is the races virulent on the common cultivars that are expected to increase because of their adaptation to these cultivars. (This is why vertical resistance is notoriously transient in effectiveness.) The reversal from the ordinary makes deeper analysis desirable.

* The analyses in Fig. 7.1 and Table 7.3 were presented by me at a Research Branch Seminar of the Canada Department of Agriculture held at Ottawa in May 1973.

Table 7.3 analyzes other data of Green (1971), and includes all recorded races, both within and without the 15B group. It covers the years 1965, 1967, 1968, and 1969, the year 1966 being excluded because Green (in his Table 10) does not give information about virulence on the gene Sr_{9d}. The table starts by showing the percentage of isolates virulent on Sr_5, Sr_6, and Sr_{9d}, each singly. This information comes directly from Green. Then follows the percentage expected to be virulent on two genes combined, if the virulences occur independently of one another and are selectively neutral. For example 81.0% of the isolates in 1965 were virulent on Sr_5 and 65.1% on Sr_{9d}, whence $65.1 \times 0.810 = 52.7\%$ are expected to be virulent on Sr_5 and Sr_{9d} combined. By analyzing the data Green gives about races, the actual percentage virulent on both Sr_5 and Sr_{9d} was found to be 56.1. The difference between 52.7 and 56.1% is statistically insignificant; and in all 4 years the percentage found in the survey tallies well with the percentage expected. So too with the combination of virulences on Sr_5 and Sr_6 the percentage found tallies well with the percentage expected. It is the combination of virulences on Sr_6 and Sr_{9d} that differs widely.

TABLE 7.3

Percent of Total Isolates of *Puccinia graminis tritici* in Canada Virulent on Resistance Genes Sr_5, Sr_6, and Sr_{9d}, Singly and in Combination, in the Years 1965, 1967, 1968, and 1969

Resistance gene	Year [a]			
	1965	1967	1968	1969
Sr_5 [b]	81.0	98.5	96.5	99.4
Sr_6 [b]	10.4	13.5	18.8	8.1
Sr_{9d} [b]	65.1	80.2	79.2	91.3
$Sr_5 + Sr_6$ expected [c]	8.4	13.3	18.1	8.1
$Sr_5 + Sr_6$ found	7.5	13.5	18.8	8.1
$Sr_5 + Sr_{9d}$ expected [c]	52.7	79.0	76.4	90.8
$Sr_5 + Sr_{9d}$ found [d]	56.1	78.7	74.2	90.7
$Sr_6 + Sr_{9d}$ expected [c]	6.8	10.8	14.9	7.4
$Sr_6 + Sr_{9d}$ found [d]	0	0	0	0
$Sr_5 + Sr_6 + Sr_{9d}$ expected [c]	5.5	10.6	14.4	7.4
$Sr_5 + Sr_6 + Sr_{9d}$ found [d]	0	0	0	0

[a] The total number of isolates was 373, 207, 202, and 172 for the four years, respectively.

[b] Data of Green (1971).

[c] Expected on the assumption that the virulences on the resistance genes occur independently of each other in the fungus population and are selectively neutral.

[d] Excluding race C25 which was found in 1965 and 1968, but not in 1967 and 1969.

In the years of Table 7.3 no isolates with this combination of virulences were found during the survey, a highly significant difference statistically from what was expected. Inevitably also the percentage of isolates found virulent on the triple combination Sr_5, Sr_6, and Sr_{9d} was also very significantly less than the percentage expected.

In Canada *Puccinia graminis tritici* did not exist in substantial amounts with the double virulence on Sr_6 and Sr_{9d} needed to attack Selkirk. The usual adaptation of pathogen to host—the adaptation that has ruined so much vertical resistance—did not occur. If the pathogen had adapted itself, the frequency of double virulence on Sr_6 and Sr_{9d} would have been greater than what was expected on the assumption of random distribution and selective neutrality. In actual fact, it was very much less. Canadian wheat was saved.

Table 7.3 confirms Fig. 7.1, and both are confirmed by the evidence from the wheat nurseries established every year across Canada. In these nurseries wheat varieties with the gene Sr_{9d} but not Sr_6 commonly became severely infected with stem rust. A variety (McMurachy) with Sr_6 but not Sr_{9d} was sometimes severely infected. But Selkirk, with both Sr_6 and Sr_{9d}, usually escaped infection or was infected only in trace amounts. The combined evidence is overwhelmingly strong that up to the present (no predictions can be made for the future) *P. graminis* has been unable to combine virulence on Sr_6 with virulence on Sr_{9d} in a phenotype adequately fit to survive in Canada, despite the great selective advantage that would accrue to it if it did.

The background is equally noteworthy. Combined virulence on Sr_6 and Sr_{9d} is common in Texas where *P. graminis* overwinters in the uredial stage. Stewart *et al.* (1970) found that in 1968 half the isolates of *P. graminis* from wheat in Texas could attack Selkirk. Somewhere on the journey northwards from Texas to Canada, some time between winter and summer, *P. graminis* sheds its double virulence on Sr_6 and Sr_{9d} and arrives with these virulences occurring separately. Adaptation is countered. In Texas virulence on Selkirk or its like is unnecessary, but occurs. In Canada virulence on Selkirk would have been highly advantageous, but failed to occur except in trace amounts. In Canada, or on the way to Canada, stabilizing selection intervened and nullified directional selection towards the appropriate combination of virulences. It is to this stabilizing selection that the safety of Selkirk wheat has been due.

We can now return to the second gene-for-gene hypothesis. It states that virulences vary in quality as shown by their fitness to survive. We have presented strong evidence for this. In Canada in combination with virulence on Sr_{9d}, virulence on Sr_5 has survived significantly better than virulence on Sr_6; and in combination with virulence on Sr_6, virulence on Sr_5 has survived significantly better than virulence on Sr_{9d}. We are concerned here only to establish the second gene-for-gene hypothesis. Other implications of the evidence, e.g., that the environment—Canada in this case—is important, will be taken up later.

To establish the hypothesis it remains to present evidence, first, that there is a causal connection between the fitness of the parasite to survive and resistance in the host, and, second, that the virulence gene itself can affect fitness to survive even when it is unnecessary. These matters concern potato blight as much as wheat stem rust. But the same arguments hold, and only wheat rust need be considered.

Causal connection is the antithesis of accidental happening. In the present context, it implies that the lack of fitness of virulence to survive when the virulence is unnecessary is no haphazard accident of creation, but is determined by the matching resistance gene. To illustrate the alternatives of causal connection and accidental happening, consider virulence of *Puccinia graminis tritici* on Selkirk wheat, or, more specifically, consider combined virulence on genes Sr_6 and Sr_{9d}. The facts given in Fig. 7.1 and Table 7.3 show that combined virulence on Sr_6 and Sr_{9d} was rare in Canada. The two races with combined virulence were 15B-3 (Can.) and 15B-5 (Can.). Green (1971) notes that these races failed to become prevalent, apparently because they were not aggressive. Were these races weak by accident or were they inevitably weak in Canada? If the weakness was just an accident, it follows that normally races with combined virulence on Sr_6 *and* Sr_{9d} would not be weak. If so, why did they survive so poorly, as Fig. 7.1 and Table 7.3 show? In any case, it is a function of statistical tests to evaluate the probability of accidents. Table 7.3 shows that double virulence on Sr_6 and Sr_{9d} is much rarer than expected; and a X^2 test shows that the probability of this being due to chance is exceedingly small. What holds for chance holds for accident, because their meanings are identical.

The evidence that double virulence on Sr_6 and Sr_{9d} survives abundantly in Texas is equally against ascribing lack of fitness in Canada to accident. It would be against all probability to assume that by accident only the least fit part of the pathogenic population survives the journey northward.

The other matter needing discussion is evidence that the virulence gene itself can affect the fitness of the pathogen to survive, i.e., that the effect is not just the result of linkage. The linkage theory has been put forward by Favret (1971), who suggested that mutation to virulence allows unfavorable linkages to show up.

Our gene-for-gene hypothesis demands that the virulence gene itself should be able to affect fitness. This is not to say that the virulence gene alone should be able to affect fitness. There is nothing in the hypothesis to require the virulence gene to act but not interact. Interactions, whether they be direct or epigenetic, are clearly permissible. Indeed, the evidence already presented about wheat stem rust suggests interaction, though not between linked genes. In so far as Favret's suggestion of linkage implies an indirect action of virulence genes, there is nothing against it except lack of evidence. In so far as the suggestion of linkage might imply that only the less fit part of the population has been able to mutate from avirulence to virulence, we dismiss it as highly improbable. There is no evidence for such linkage; and there is no point in calling on a improbable explanation when more probable explanations are available. A more probable suggestion is made in the next section.

7.7 A THEORY OF LEAKY MUTANTS AND MISSENSE MUTATION

There is substantial evidence reviewed by Van der Plank (1968) that mutation to virulence in *Phytophthora infestans* is reversible. The simplest possible explanation would be that one amino acid side chain in the coded protein is replaced by another. As we saw in Section 7.5, when virulence is unnecessary mutation to virulence on the potato resistance gene R_1 has a greater effect on the fitness of *P. infestans* than mutation to virulence on R_4. The suggestion is that mutation to virulence on R_1 involves the substitution of one amino acid by another with a dissimilar side chain, whereas mutation to virulence on R_4 involves substitution by a more similar amino acid that has little effect on the configuration of the coded protein. Avirulence and virulence on R_4 are about equally fit to survive in the absence of R_4 because the coded proteins of avirulence and virulence differ little in configuration. But, it is suggested, the coded proteins of avirulence and virulence on R_1 differ more.

So too it is suggested that in *Puccinia graminis tritici* mutation to virulence on Sr_6 and Sr_{9d} involves the substitution of amino acids by others with more different side chains than mutation to virulence on Sr_5 does.

Our theory is simply a theory of leaky mutants, with varying amounts of leakiness. It is generally accepted in genetics that most mutants are leaky, so the theory follows a conventional pattern.

In alternative genetic terminology, the theory is that mutations to virulence are missense mutations, and that the substitution is more conservative during mutation to virulence on the gene $R4$ than during mutation to virulence on $R1$. The second gene-for-gene hypothesis recognizes that there are degrees of conservatism.

7.8   AN ESSENTIAL POSTULATE. SOME LIGHT ON DOMINANCE AND ALLELISM

What was written in the previous section implies a simple postulate: Genes for avirulence have functions other than those manifested in the presence of resistance genes in the host; and in the absence of resistance genes these other functions are beneficial to the pathogen. We call a gene an avirulence gene simply because avirulence is its only attribute we are aware of. But the term, avirulence gene, is really a misnomer. The gene is beneficial to the pathogen in the absence of resistance genes in the host, and lethal only in their presence.

Consider *Phytophthora infestans*. It struck the potato fields of Europe in 1845. At that time fields of potatoes, being pure *Solanum tuberosum,* had no known R genes, and (outside of a few greenhouses and botanic gardens) the fungus in Europe did not meet any known R genes until about 1935 when the gene R_1 was commercially introduced from *S. demissum.* During those 90 years there was nothing to indicate the existence of avirulence. Nevertheless, during all those years there existed in *P. infestans* a segment of nucleic acid, which we later came to call the gene for avirulence on resistance gene R_1. Moreover, this segment was universally, or almost universally, present in the fungus, because when the gene R_1 was first introduced into the potato it gave practically complete protection. Our postulate states that during those 90 years (and thereafter in potato fields without R genes) the miscalled avirulence gene was carrying out a beneficial function in the pathogen.

We infer the function was beneficial. There is a consensus among population geneticists that if a gene is abundant or universal in a population, it has a use; or, conversely, that a useless gene tends to drop out of a population. Avirulence on potato gene R_1 was universally, or almost universally, present in *P. infestans*; accordingly, we infer it was useful. It seems improbable that a gene would be universally present if its only function was to be at times lethal to the phenotype; and doubly improbable that this gene would be both lethal and dominant, as avirulence usually is over virulence, when parasite and host are incompatible.

Interpreted by this postulate, the substance of the previous section is this. With respect to gene R_4, the protein coded by the gene for avirulence and that coded by the gene for virulence differ little, and in the absence of the resistance gene in the host the one can substitute for the other in carrying out the beneficial functions. But with respect to the gene R_1, the protein coded by the virulence gene is inferior and less able to take over the beneficial functions of the protein coded by the avirulence gene.

It is unfortunate that custom makes us talk of avirulence alleles, as if avirulence were the alleles' real function. The essential beneficial function is obscured by reference to avirulence and all it connotes. Perhaps we should talk of a primary useful role of the gene and a secondary harmful role manifested only in situations of host–parasite incompatibility. Confusion over the two roles has led to confusion about dominance. In its secondary role avirulence is usually dominant over virulence. From this Kiyosawa (1971) argues that virulence alleles do not produce enzymes or other active substances. It may be true that virulence alleles produce no substances active within the confines of the secondary role, i.e., no substances active in the production of whatever the particular host–parasite incompatibility reactions may be. But to state that virulence alleles produce no active substances is to ignore their primary role. No observations have ever been made about dominance in the primary role. Indeed, discussions about dominance in the primary role would often seem to be unreal. For example, if we accept that in their primary role the products of avirulence and virulence on the blight resistance gene R_4 are apparently interchangeable and more or less equal in their contributions to fitness, the concept of the dominance of avirulence here would lapse, because the distinction between avirulence and virulence would lapse.

The facts of allelism illuminate the postulate of a primary useful role of avirulence. Resistance genes in the host are commonly allelic, thus reducing the number of loci involved. But, so far as is known, virulence genes in the pathogen are not allelic, thus extending to the full the number of loci. This is what one might expect if these loci in the pathogen have a primary useful role. These loci in the pathogen may be even more numerous than we are aware of. See Section 7.11.

To judge by the evidence about antigens in Section 7.20, the primary useful role of the avirulence gene may be to code for a protein similar to, or identical with, a protein of the host plant, this host protein being coded for by the gene for susceptibility.

7.9 THE COMMONNESS OF WEAK RESISTANCE GENES

The general genetic rule is that point mutants with small phenotypic effects are commoner than those with large effects. Mutations from avirulence to virulence are more likely to have little effect on the fitness of the pathogen to survive in the absence of host–pathogen incompatibility than they are to have a large effect. In the absence of host–pathogen interactions, i.e., in their primary useful roles, avirulence and virulence are more likely to differ little than much in their effect on fitness. Mutation to virulence is likely to be conservative.

All this has a bearing on the use of resistance genes to control disease. Disease is controlled by resistance genes only if avirulence occurs more frequently than virulence. Put differently, there must be stabilizing selection in favor of avirulence to counterbalance directional selection towards virulence. The necessary condition for this, that avirulence must be considerably more fit to survive than virulence, is, on what has just been said, likely to be the exception rather than the rule.

Looking at resistance genes from the plant breeder's point of view, Van der Plank (1968) classified them as strong or weak. The classification puts the emphasis on the resistance gene, because that is what the plant breeder manipulates, but is really a classification of the matching virulence gene. Potato blight resistance gene R_4 is weak for the breeder because it is practically useless to him for controlling blight; and it is practically useless because virulence on this gene is as fit to survive as avirulence even when virulence is unnecessary. The gene R_1 is stronger and has been more useful to breeders because virulence on it is less fit to survive when it is unnecessary.

One expects weak resistance genes to be commoner than strong, and this is what one finds. With diseases that have been well studied, such as the cereal rusts or potato blight, a plethora of resistance genes have been revealed that seem almost useless to plant breeders.

Gene strength is not an absolute character. It is influenced by geography and environment, probably reflecting climate, and may perhaps be influenced by genetic interaction.

7.10   An Unsolved Geographical Problem

In *Puccinia graminis tritici* combined virulence on genes Sr_6 and Sr_{9d} is rare in Canada, as Table 7.3 shows. But it is relatively common in Texas where Stewart *et al.* (1970) found half the isolates to have this combined virulence, and in the Yaqui Valley of Mexico where Roelfs and McVey (1972) found the combined virulence in almost all isolates.

In Canada *P. graminis tritici* does not survive the winter except in now rare barberry bushes. Essentially it is a hit-and-run raider. It comes up from the south in early summer, and departs southward at the end of the season. It overwinters mainly in Texas and near by. Figure 7.2 illustrates the situation as it existed when Selkirk was the common Canadian prairie wheat variety after it had stopped the stem rust epidemic of the early 1950's. Somewhere between Texas and Canada combined virulence on Sr_6 and Sr_{9d} drops out; and for more than 20 years Canadian wheat varieties with these two resistance genes have escaped severe stem rusting.

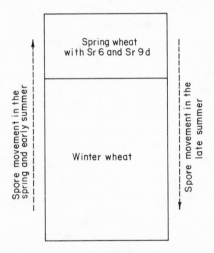

Fig. 7.2   A simplified representation of the movement of the wheat stem rust fungus to and from the red spring wheat area of North America, as it was when Selkirk was the common spring wheat. Combined virulence on Sr_6 and Sr_{9d} is (to judge by recent surveys) common in the south of the winter wheat area, but rare in the spring wheat area of the north. For this reason Sr_6 and Sr_{9d}, combined, have remained effective in spring wheat; and Sr_6 which was added to Sr_{9d} by backcrossing saved Canadian wheat from the stem rust epidemics of the early 1950's.

Temperature may be involved. Canadian wheat fields in summer are substantially warmer than Texas fields in winter. And *P. graminis tritici* journeys northward mostly at warm temperatures. It hits wheat fields mainly as they approach maturity at increasing temperature; and as the population of the pathogen moves stepwise through Texas, Oklahoma, Kansas, Nebraska, and onwards the bulk of it at any time will be found in wheat nearing ripeness and at the warm temperatures that prevail during the ripening stages of the wheat season.

Van der Plank (1968) analyzed data published by Katsuya and Green (1967) and showed that at the low temperature of 15°C an isolate of race 15*B* of *P. graminis tritici* survived well in comparison with an isolate of race 56, but at the high temperature of 25°C the isolate of race 15B was considerably less fit to survive than that of race 56. Historically the main difference between race 15*B* and race 56 has been the former's ability to attack wheat with the gene Sr_{9d}; and there is a hint in this that virulence on Sr_{9d} survives poorly at high temperatures. If the hint is correct, one would expect virulence on Sr_{9d} to fare badly during the northward migration of the pathogen's population. But where does virulence on Sr_6 enter the story? Uncombined with virulence on Sr_6, virulence on Sr_{9d} reaches Canada in substantial amounts, and wheats with Sr_{9d} but not Sr_6 are open to attack. To ignore Sr_6 is to play Hamlet without the Prince, because as a matter of history it was Sr_6 that saved Canadian wheat from the stem rust epidemic of the early 1950's. At that time Sr_{9d} was common in Canadian wheat, and only when Sr_6 was added to Sr_{9d} by backcrossing did the epidemic stop.

7.11 FOUR ARBITRARY CATEGORIES OF ADAPTATION IN THE PATHOGEN TO THE HOST. ABUNDANT PREEXISTING VIRULENCE. VIRULENCE BY ADAPTATION. RESTRICTED VIRULENCE. FORBIDDEN VIRULENCE

Abundant preexisting virulence is exemplified by virulence in *Phytophthora infestans* on resistance gene R_4. Virulence preexists abundantly even in wild-type populations that have not been on host plants with R_4. Virulence exists abundantly because of its fitness to survive even when the pathogen has no need to adapt itself to resistance in the host.

Virulence on R_4 coexists with avirulence in wild-type populations. Possibly the failure of one allele to displace the other stems from a fluctuating environment that favors first the one and then the other. Alternatively it may stem from what seems to be an abnormally high mutation rate from avirulence to virulence, forward and backward.

It is possible that there are examples more extreme than that of virulence in *P. infestans* on R_4. There could conceivably be virulence that, even in the absence of matching resistance in the host, exists so abundantly as to be almost universal. If so, we are unlikely to know of it. It needs avirulence to identify a resistance gene, and a resistance gene to identify virulence; and occasional avirulence that arose by mutation would be difficult to capture. This matter is of no interest to the practical plant breeder. Nevertheless, it is worth remembering that the number of loci concerned in virulence may be larger than we are likely to be aware of. In terms of the postulate in Section 7.8 it could be argued that the larger the number the greater the advantage to the pathogen might be.

It has already been noted in Section 7.5 that with virulence like virulence on R_4 very little vertical resistance is available to plant breeders. With virulence of this sort there is not even the initial boom in the boom-and-bust sequence so notorious in vertical resistance. The resistance gene is practically useless from the start.

The second category, virulence by adaptation, is exemplified by virulence in *P. infestans* on R_1. Virulence existing before the use of R_1 in potato fields was scarce. But when potatoes with R_1 were introduced and became widely grown, virulence became abundant as an adaptation to the resistance. Where potatoes with the gene R_1 are widely grown the adaptation is practically complete, and the potato fields must be protected with a fungicide as if the gene did not exist.

With virulence in this category, vertical resistance follows a boom-and-bust sequence. The scarcity of preexisting virulence allows the boom; adaptation causes the bust.

The third category, restricted virulence, is exemplified by combined virulence in *Puccinia graminis tritici* on Sr_6 and Sr_{9d} in wheat in Canada. Virulence has appeared; but up to the present it has remained restricted.

While virulence is restricted, vertical resistance remains stable. The boom-and-bust sequence stays at the boom.

Restricted adaptation is not necessarily an inherent property of the genes concerned. Environment and life histories are relevant. The adaptation of combined virulence on Sr_6 and Sr_{9d} has not been restricted in Mexico or Texas; and it may well be that the restriction in Canada derives from a life history that keeps *P. graminis tritici* out of Canada most of the year. In nonobligate parasites conditions during the saprophytic phase may restrict adaptation; and if virulence lowers fitness to survive saprophytically, adaptation will be restricted.

A tendency to adapt is universal in all living beings. It can be taken as axiomatic that pathogens will tend to acquire the broadest spectrum of virulence that will adapt them to hosts. That is, there will be directional selection in favor of virulence. Against this will operate stabilizing selection if virulence reduces fitness to survive when it is unnecessary. The balance is determined not only by the quality of the genes as defined in the second gene-for-gene hypothesis but also by circumstance and life history. If adaptation dominates, virulence falls in the second category; if stabilizing selection dominates, virulence falls in the third category.

Because so many unknown factors affect adaptation and stabilizing selection one can read past history without being sure of the future. Combined virulence on Sr_6 and Sr_{9d} in Canada has been rare, but because we do not know why it was rare we cannot exclude the possibility of greater adaptation to come. The year 1960 gives a hint of potential danger. In that year, Fig. 7.1 shows, there was a small flare of combined virulence, not enough to do serious harm but enough to show that we do not know what is going on. Strangely, 1960 saw the nadir of race $15B$ in two decades, a reminder of our ignorance and a clue perhaps of negative correlations.

The fourth category, forbidden virulence, concerns immunity. Adaptation is excluded. Immunity is discussed near the end of this chapter.

Division into four categories is arbitrary. There is probably a continuum from one extreme to the other—from abundant preexisting virulence, even to the extent where resistance is cryptic, to virulence that is forbidden in any environment or during any part of a life history.

Of the categories, immunity is the commonest, followed by the other extreme of preexisting virulence. The intermediate categories—virulence by adaptation and restricted virulence—seem to be relatively rare, but are the categories important in vertical resistance.

7.12   THE NEED TO STUDY POPULATIONS IN ORDER TO ESTIMATE FITNESS. THE ERROR OF TYPIFYING POPULATIONS BY SINGLE ISOLATES

The topic of this chapter has been one of population genetics, and special interest has centered around the frequency of virulence or avirulence. The data were derived from a number of isolates drawn from the population. Thus in Section 7.5 the data of Frandsen in Germany and Graham in Canada were used to study in populations of *Phytoph-thora infestans* the frequency of virulence on the resistance genes R_1 and R_4. Statistical tests then showed that the number of isolates was enough to establish the conclusions with a high degree of probability, i.e., they showed that the isolates adequately represented the population.

Individual differences exist in populations of pathogens as in most other populations; and conclusions based on single isolates can mislead. Thurston (1961) and Thurston and Eide (1952, 1953) set out to compare the fitness of races 0 and 1 of *P. infestans* to survive on potatoes without R_1. They used isolates. After several passages through the potatoes, race 0 predominated. But another experiment reversed the result, and race 1 predominated. Thurston (1961) concluded correctly that although, in general, the unnecessary virulence on R_1 reduced fitness it is possible to find individual isolates without unnecessary virulence that are relatively unfit. In other words there are constitutional factors other than virulence that also affect fitness. But it would have been more correct for him to have concluded that the experiments were scarcely worth doing in the first place, because they confused an isolate with a population.

Tom Johnson is 6 ft tall. Tom Johnson is an American. Americans are 6 ft tall. Typifying Americans by an American makes no sense. But it is no more nonsensical than trying to typify a pathogenic population by a single isolate. The fitness of a pathogen is polygenic; it is determined by the whole genome and not just by genes for virulence or avirulence. Within the population variation between individuals with the same virulence is inevitable, and this must be allowed for in assessments of fitness. An illustration of variation between isolates will be found in Table 7.4.

TABLE 7.4

Combined Analysis of Variance of Two Tests of Four Isolates of *P. infestans* on Three Potato Varieties [a]

	Degrees of freedom	Mean square	P
Isolates	3	563.7	<0.001
Varieties	2	919.0	<0.001
Isolates × varieties	6	23.5	
Tests	1	1230.0	
Error	132	19.6	

[a] Data of Paxman (1963).

7.13   HORIZONTAL RESISTANCE. ABSENCE OF DIFFERENTIAL
INTERACTION

Horizontal resistance is defined as resistance in which there are no differential interactions between races of the pathogen and varieties of the host. Two sets of evidence about potato blight illustrate the concept.

Paxman (1963) set out to determine whether races of *Phytophthora infestans* would become specially adapted to a potato variety if they were grown continuously on it. He used varieties without R genes. He obtained an isolate 30RS from a naturally infected tuber of the variety Red Skin, another isolate 31KP from a naturally infected tuber of Kerr's Pink, and an isolate 32KE from a naturally infected tuber of the variety King Edward. These he subcultured, each on its original cultivar, i.e., he subcultured isolate 30RS on tubers of Red Skin, 31KP on tubers of Kerr's Pink, and 32KE on tubers of King Edward. He started his tests after 90 cycles of subculturing. That is, the isolate 30RS had been on the variety Red Skin for 90 cycles plus the unknown number of cycles the fungus had been on Red Skin in the field before it was isolated. He used as his criterion the rate of spread of each of the three isolates in tubers not only of the variety of origin, e.g., isolate 30RS in Red Skin, but also in tubers of the other two varieties, e.g., isolate 30RS in tubers of Kerr's Pink and King Edward as well. He also used a fourth isolate of unspecified origin but cultured on the variety Majestic. His results, presented in Table 7.4, show a highly significant difference between isolates. That is, the isolates differed significantly in their ability to spread through tuber tissue. There was also a highly significant difference between varieties. That is, varieties differed significantly in the resistance they offered to mycelium spreading through tuber tissue. But—and this is the gist of Paxman's results—there was no evidence for an isolate × variety interaction. The pathogenicity of the isolates was uniformly spread over the host varieties; and the resistance of the host varieties was uniformly spread against the isolates. Resistance was wholly nondifferential.

Van der Plank (1971) analyzed numerical assessments of resistance to potato blight in the Netherlands. Data were available on ten potato varieties that had been on the official lists continuously during the 30 years, 1938 to 1968. None of these varieties had an R gene. They varied from one another in resistance. Some were very susceptible, some moderately resistant. During the 30 years there had been some great changes. There had been a great change of varieties, only the ten surviving the change. There had been a great change in virulence of the pathogen, virulence on gene R_1 in particular having increased considerably in abundance. Yet, despite these changes, resistance both in the foliage and the tubers had remained substantially stable. The assessments of resistance made in 1968 were substantially the same as those made in 1938, with little change in the ranking of the varieties.

The definition rules out the possibility of any gene-for-gene relation in horizontal resistance. To say that there is no differential interaction between varieties of the host and races of the pathogen is to say that varieties of the host do not have their "own" races of the pathogen, and races of the pathogen do not have their "own" varieties of the host. This being so, no gene-for-gene relationship is conceivable. Indeed, it would be as good a definition of horizontal resistance as any to say that it is resistance without possible gene-for-gene relations.

7.14 ANTHRACNOSE OF BEANS. EXAMPLES OF PHYSIOLOGICAL REACTIONS INVOLVED IN VERTICAL AND HORIZONTAL RESISTANCE

For further illustration consider the disease of *Phaseolus vulgaris* caused by *Colletotrichum lindemuthianum*. The fungus occurs in several distinct races—the alpha, beta, gamma, and delta races—and there is a marked differential interaction between cultivars of the bean and races of the fungus. See Table 7.5. In other words, there is clear verti-

TABLE 7.5

Response of *Phaseolus vulgaris* Cultivars to the Beta, Gamma, and Delta Races of *Colletotrichum lindemuthianum* [a]

Cultivar	Response to race [b]		
	Beta	Gamma	Delta
Topcrop	S	R	S
Wade	S	R	R
Harvester	S	R	S
Improved Tendergreen	S	R	S
Perry Marrow	R	S	S
White Marrowfat	R	S	S
Tennessee Green Pod	R	R	S

[a] Data of Berard *et al.* (1973).

[b] S: susceptible; R: resistant.

cal resistance. But horizontal resistance is universally present in all host plants to some extent or other (see the next section). So the resistance of the bean host is a mixture of the two sorts of resistance. This is clearly reflected in studies that have been made.

Berard *et al.* (1973) studied the protection given by diffusates from incompatible interactions of host and pathogen, i.e., by diffusates obtained after treatment of the host by avirulent cultures of the pathogen. Their study is one of vertical resistance, because it is a study of differential interactions in resistance. Drops of suspensions of conidia of isolates of the appropriate pathogenic races were applied to the hypocotyls of bean seedlings, left there for 60 hours, collected, sterilized, and concentrated. The concentrated solution is referred to as the diffusate. They used isolates of the beta, gamma, and delta races. The cultivars Topcrop, Wade, Harvester, and Improved Tendergreen are all susceptible to the beta race but resistant to the gamma race. See Table 7.5. A diffusate obtained by treating Topcrop with the (avirulent) gamma race protected these four cultivars from the (virulent) beta race. The cultivars Perry Marrow and White Marrowfat are resistant to the beta race but susceptible to the gamma race. A diffusate obtained by treating Perry Marrow with the beta race protected these two cultivars from the gamma race. But a diffusate obtained by treating Topcrop with the gamma race did not protect Perry Marrow from this race. The cultivar Tennessee Green Pod is resistant to the beta and gamma races but not the delta race. Diffusates from either the beta or gamma races on this cultivar protected it from the delta race. But Perry Marrow was protected from the gamma race only by the diffusate from the beta race on Tennessee Green Pod, but not the diffusate from the gamma race on this cultivar. These and other findings show that there are at least two different factors involved in the interactions. But what is being illustrated here is that the interactions are markedly differential, as is required by factors involved in vertical resistance.

Of quite another sort are the findings of Anderson and Albersheim (1972). They illustrate a process in horizontal resistance. Anderson and Albersheim studied the endopolygalacturonases secreted by isolates of the alpha, gamma, and delta races of *C. lindemuthianum* and inhibitor proteins associated with the cell walls of the bean cultivars Red Kidney, Small White, and Pinto. These cultivars are differentially resistant to the races. Red Kidney is resistant to the alpha race, partially resistant to the delta race, and susceptible to the gamma race. Small White is resistant to the gamma race, but susceptible to the alpha and delta races. Pinto is susceptible to the three races. But this differential

interaction is not reflected in the relation between the endopolygalacturonases of the races and the inhibitor proteins of the cultivars. The
relation is uniform (nondifferential), and consistent with its being part
of the process of horizontal resistance.

As background, the ability of the pectic degrading enzymes of pathogens to macerate host tissue and kill cells indicates their importance in
the infection process. The literature has been reviewed by Wood
(1967), Bateman and Millar (1966), and Albersheim *et al.* (1969). A
purified preparation of the endopolygalacturonase from the alpha race
of *C. lindemuthianum* is able to degrade all walls with the removal of
considerable amounts of pectic material (English *et al.*, 1972). Cell
walls must first be treated with this endopolygalacturonase before they
can be further degraded by other enzymes such as cellulase, arabinosi-
dase, and pronase (English *et al.*, 1972; Jones *et al.*, 1972). The endopolygalacturonase is clearly part of the pathogenicity apparatus of the
fungus, and the genes controlling its production are pathogenicity genes.
To turn from the pathogen to the host, it has been established that the
cell walls of plants contain a protein which can specifically and effec-
tively inhibit polygalacturonases of fungal origin (Albersheim and Anderson, 1971). The degradation of tomato cell walls by culture filtrates
of *Fusarium oxysporum* f. sp. *lycopersici* is prevented by the addition
of this protein inhibitor (Jones *et al.*, 1972). The protein is clearly part
of the resistance mechanism of the host, and the genes controlling its
production are resistance genes.

Anderson and Albersheim (1972) grew the three races of *C. linde-*
muthianum—the alpha, gamma, and delta races—on citrus pectin as a
carbon source, and found that the endopolygalacturonases secreted by
them were indistinguishable from one another. They also purified the
protein inhibitors from the hypocotyls of Red Kidney and Pinto beans.
These inhibitors were found to be indistinguishable, and equally able to
inhibit the endopolygalucturonases from the three races of *C. lindemu-*
thianum. Also, the purified endopolygalacturonases from the three races
equally competed for the inhibitor. Because the inhibitor was spread
uniformly against the endopolygalacturonases, without detectable differential interaction, resistance given by the inhibitors is horizontal, and
inhibition of endopolygalacturonases by cell wall proteins can be taken
as one of the thousands of processes that go to make up horizontal resistance.

For emphasis let us repeat two points so that there can be no misunderstanding. Cell walls must be attacked by endopolygalacturonase before they can be further degraded by other enzymes. The genes in the

fungus that control the synthesis of endopolygalacturonase are therefore unquestionably genes for pathogenicity. They are as much genes for pathogenicity as are the genes for virulence discussed earlier in this chapter. So too the genes controlling the production of enzymes that further degrade the cell walls are also all genes for pathogenicity. Cell walls contain a protein that inhibits the polygalacturonase. The genes concerned in the synthesis of this protein are genes for resistance, as truly as the Sr and R genes discussed earlier are genes for resistance. No progress in understanding horizontal resistance can be made unless this elementary point is taken.

7.15   THE COMMONNESS OF HORIZONTAL RESISTANCE. HORIZONTAL RESISTANCE THROUGH NORMAL METABOLIC PROCESSES

Let us return to Paxman's experiments on the invasion of potato tuber tissues by *Phytophthora infestans* and consider the obvious. *Phytophthora infestans* changed healthy tuber tissue into diseased tuber tissue, and thereby earned a living. To do so it needed enzymes: pectinases, pectin–methylases, cellulases, proteinases, and the like, all in yet unmeasured diversity. These enzymes are part of the parasitic process; and the genes that control them are *ipso facto* genes for pathogenicity. The tubers on their side had pectins, celluloses, proteins, and other constituents which had to be broken down by the pathogen. They were barriers to infection. Barriers are synonymous with resistance, and the genes that controlled the laying down of these barriers are *ipso facto* genes for resistance.

Consider some other experiments with potato blight. Grainger (1956, 1957, 1959) worked with potato varieties that had no R genes, and found a close correlation between susceptibility and the ratio of total carbohydrate to the residual dry weight of the shoot. It seems that resistance is greatest when the amount of "spare" carbohydrate in the host is least, i.e., *P. infestans* thrives best on a diet high in carbohydrate. Grainger's results with potato blight have been examined recently by Warren *et al.* (1973). They confirmed that there is a positive correlation between susceptibility and the (total carbohydrate)/(residual dry matter) ratio, when conditions are natural. But they failed in attempts to extend the correlation to results obtained in artificial conditions. Thus, sugar injections into the plant increased susceptibility to blight, which accords with Grainger's hypothesis, but surprisingly failed to increase the (total carbohydrate)/(residual dry matter) ratio. The ratio may be a blunt and arbitrary tool; but at least it seems clear, first, that the pathogen's aggressiveness varies with its food supply, and, second, that carbohydrate is one of the foods of *P. infestans*.

We need not concern ourselves much about the accuracy in detail of Grainger's conclusion. It is obvious that parasites depend on available food, and available food depends in turn on a multiplicity of factors. Genes controlling these factors are *ipso facto* genes for resistance or susceptibility. This point was missed by Grainger himself, who wrote that the host's physiological ability to permit fungal attack is quite different from genetically controlled resistance or susceptibility. Genetic control of any physiological, biochemical, anatomical, or morphological detail of the host that affects resistance to disease is genetic control of resistance or susceptibility.

It is no part of the theory of genetic control of disease to require that resistance be controlled by the direct action of the gene or to assume that genes for resistance have no functions other than to provide resistance.

The literature of the relation between the carbohydrate content of plants and proneness to attack by pathogens was reviewed by Horsfall and Dimond (1957) and Wood (1967). Some diseases are promoted, others retarded by high sugar content. We are not concerned here with these details but with the conclusion that genes that control the pathogen's food can be genes for resistance or susceptibility.

Examples in this section have been drawn from potato blight because of the evidence (see Section 7.13) that resistance to blight in potatoes without *R* genes is mostly or wholly horizontal. The examples are in all probability examples of horizontal resistance.

To return to Paxman's experiments discussed in Section 7.13, the potato variety Kerr's Pink is more resistant than the variety King Edward. The absence of a differential interaction—the fact that resistance is horizontal—suggests that Kerr's Pink has no defenses not also available, though weaker in King Edward. Some isolates of *P. infestans* were more aggressive than others. Resistance being horizontal suggests that the more aggressive isolates used no enzymes not also used, though in smaller amounts or less efficiently, by the less aggressive isolates.

It is difficult to conceive of any host–pathogen combination in which the host has no barriers, physical or chemical, to infection. There is almost certainly no such thing as zero horizontal resistance. (This does not imply that the resistance is necessarily enough to satisfy the plant breeder.) But there may be zero vertical resistance; resistance to blight in potatoes without R genes may well exemplify this.

Many, perhaps all, genes for horizontal resistance exist independently of the need for resistance. They exist in geographical areas from which the pathogen is absent. They exist in healthy plants not threatened by disease. When the need arises and plants are subjected to disease, horizontal resistance can be enhanced by gene changes, but the enhancement need not involve any new form of resistance. Resistance, e.g., could be enhanced by gene duplication (see the next section). There is no evidence yet to show that horizontal resistance is created, as distinct from enhanced, to meet the threat of disease.

7.16 Vertical versus Horizontal Resistance. New Genes versus Old Genes. Oligogenic versus Polygenic Resistance. Gene Diversification versus Gene Duplication. A Molecular Theory of Horizontal Resistance

In agriculture the quest for vertical resistance has almost invariably been a quest for "new" genes, that is, genes new to the environment. Potato breeders in Europe went to Mexico to get R genes from *Solanum demissum* in order to provide resistance to blight. North American wheat breeders got Sr genes for resistance to stem rust from Kenya. Tomato breeders got resistance to fusarium wilt from *Lycopersicon pimpinellifolium*. International collections of wheat, oats, potatoes, and other important crops are maintained as pools of new genes for, *inter alia,* vertical resistance. Projects are running to seek new resistance by mutagenesis. Interspecific and intergeneric crosses are made to bring into crop plants genes from their wild relatives.

Horizontal resistance, on the other hand, has been found within existing crops, and to judge by what has been said in the preceding sections is likely to be furnished by host processes as old as the hills.

The distinction between new and old processes is relevant. Pathogens have great powers for adaptation. Block them with a new resistance gene, and they are likely to adapt themselves to it. Resistance is ended as new adapted virulence arises. That is the story of the failures of vertical resistance. But if the pathogen is confronted with processes to which it has been adapted for geological ages, further adaptation is likely to be insubstantial. That is our theory of horizontal resistance: It involves host and parasite processes that are already age old. Often at least, the parasite lives with processes it has lived with ever since the host–parasite association began.

It is worth discussing "newness." *R* genes for resistance to potato blight were new to *Solanum tuberosum* in Europe but old in *S. demissum* in Mexico. Newness is relative to the particular environment. In agriculture new genes are obtained by introduction, mutation, or hybridization. In nature newness needed for vertical resistance is maintained largely or wholly by a diversity of host genotypes within the environment. A pathogen moving from one host genotype to another makes new or renewed acquaintance with resistance genes as it moves.

Vertical resistance is oligogenic, horizontal resistance polygenic. That is another way of comparing the two sorts of resistance.

It is at present impossible to estimate accurately how many genes are concerned in resistance, but if my hypothesis is correct the number of genes of horizontal resistance to a fungus parasite must be large. The genome of an angiosperm is among the greatest yet measured, the nucleotide pairs of a haploid set of chromosomes being numbered in billions and the genes in millions. Only a fraction of these are functionally active, the fraction, on the evidence of Abrahamson *et al.* (1973), probably decreasing as the size of the genome increases; and of these functionally active genes only some are involved in the processes that affect the life of a parasite within the host. Even so, it seems not improbable that the total number of genes involved in horizontal resistance must be counted in thousands, and the description of horizontal resistance as polygenic can be taken as valid. But the number of "unfrozen" genes involved in the *variation* of horizontal resistance within a species—and it is only this variation whch is useful to plant breeders—is much smaller (see the next section); and there is no reason why the variation should not occasionally be oligogenically controlled. Although large numbers of genes are involved in horizontal resistance, it is no part of my hypothesis to attribute the stability of horizontal resistance, or the instability of vertical resistance, to numbers as such.

Vertical resistance involves gene diversity; its effects are qualitative. It probably does not involve gene duplication, except in the form of homozygosity when resistance is recessive; then allele duplication is necessary to carry diversity from the genotype to the phenotype. Horizontal resistance involves gene dose; its effects are quantitative. Probably, horizontal resistance or susceptibility is governed largely by gene duplication. Gene duplication retains oldness. During man's short spell of breeding crop plants for horizontal resistance there might not have been time enough for duplication to have done more than bring together a greater number of appropriate alleles without enlarging the number of loci on which they occur. But in nature age-long selection might well have accumulated tandem or displaced duplications in additional loci. Equally (because horizontal resistance and susceptibility are regarded as opposites in terms of gene dose), horizontal resistance or susceptibility could be affected by deletions.

Horizontal resistance can at times be altered by mutation. A reasonably sure example is the increased horizontal resistance to blight that accompanies the later maturity of potatoes that have "bolted." The genetics of bolting has yet to be satisfactorily established—it might involve duplication—but it would be wise at this stage to assume that, in addition, horizontal resistance could be altered by mutations that do not necessarily involve duplication or deletion. These mutations could still fit the suggestion that horizontal resistance is concerned with old, familiar host–pathogen processes, e.g., that bolted potatoes present *P. infestans* with no problems not also present though to a lesser degree in normal potatoes. It is necessary only to assume that mutation brings in no secondary effects that require special adaptation by the pathogen. For example, if by way of illustration one adopts Grainger's theory (see Section 7.15) that resistance to blight in potatoes is determined by the amount of "spare" carbohydrate, it is necessary only to assume that bolting in potatoes affects the amount of spare carbohydrate at any given time, without introducing new sugars or presenting *P. infestans* with new problems that its pathogenic processes are not already adapted to cope with.

Considered chemically, horizontal resistance or susceptibility is largely or wholly determined by the dose of each relevant enzyme and on how reactions are switched on and off. Diversification, as distinct from dose, probably plays little or no part; and mutants, if they occur, have no effects qualitatively new to the pathogen. Such a theory would seem to fit the known facts.

7.17 THE INEQUALITY OF HORIZONTAL RESISTANCE GENES

When inheritance is polygenic the larger the number of genes involved, the smaller is the chance of obtaining a phenotype at a given distance from the mean phenotype. This proposition in genetics is what this section is about.

If horizontal resistance were governed by a hundred thousand genes all contributing equally to variation, there would be practically no chance of substantially improving by selection the highest level of resistance now available. But if it were governed by only six genes, rapid change would probably be possible. It may well be that a hundred thousand genes are involved. The saving feature is that they are not equally involved. Most genes make only minute contributions to variation in resistance or make no contribution to variation at all.

Consider blight in potatoes without R genes, because resistance is essentially horizontal. When the variety Katahdin is self-pollinated and the progeny raised, the seedlings differ considerably in resistance. The resistance is closely coupled with late maturity, and early maturing progeny are almost invariably very susceptible. The great difference among the progeny in lateness and resistance shows that a few genes have a major share in determining these qualities. Within each maturity class there are also differences; and differences of this sort have been used by plant breeders to achieve substantial increases of resistance by selection. Presumably here, too, the number of genes greatly involved is small.

Possibly overriding everything and limiting the number of genes available for use in breeding for horizontal resistance is the restriction on variation and segregation that follows an evolutionary "freeze" on entire essential systems. For illustration we can use Grainger's theory (Section 7.15) that resistance to *Phytophthora infestans* in the potato is determined by the amount of "spare" carbohydrate. The amount of spare carbohydrate is determined by the balance between photosynthesis on the one hand and respiration and growth on the other. Everything—the position, size, shape, number, and functioning of stomata, for example—that affects any of these processes affects the amount of spare carbohydrate and therefore horizontal resistance. A large slice of the host's metabolism is involved, and the genes concerned in horizontal resistance must be numbered in thousands and tens of thousands. But when once a well-functioning metabolic system has been evolved, any substantial change in the kinetic property of one enzyme without simultaneous adjustments in the kinetic properties of functionally coupled enzymes would be disadvantageous. Mutations being rare and random, the probability that two or more independent gene loci would simultaneously sustain mutually acceptable mutation is vanishingly small, hence, the reason for an evolutionary freeze of the entire system. The same genes would be expected to occur in related varieties, species, and genera of the host plant; and hybridization would not produce those segregations in the progeny on which most plant breeding depends. The theory that horizontal resistance is governed by genes that have normal functions in the healthy plant has the corollary that relatively few of these genes are likely to be available for use in breeding for horizontal resistance (and of these few still fewer are likely to be limiting factors in resistance). It also has the corollary that tampering with established systems could be harmful, which brings us to the proposition discussed in the next section that excess horizontal resistance could be harmful. Our theory, then, is that countless genes contribute to horizontal resistance, which is always present; but that relatively few of these genes are not frozen and are available for the plant breeder to manipulate.

For an account of the concept of an evolutionary freeze, see Ohno (1973).

When in other sections we discuss phenomena such as heterozygosity or gene duplication in relation to horizontal resistance we refer particularly to that part of the inheritance which is not frozen.

7.18 THE QUANTITY OF HORIZONTAL RESISTANCE. THE HARMFULNESS OF EXCESS OF HORIZONTAL RESISTANCE

How much horizontal resistance is available? Is it enough to be worth breeding for? Does it play a substantial part in host–pathogen ecology? The difficulty about answering these questions is that the respective contributions of horizontal and vertical resistance have still to be sorted out, except perhaps with potato blight and a few other diseases. We shall discuss in detail only three deseases: two rust diseases of maize and potato blight.

Maize has both vertical and horizontal resistance to *Puccinia sorghi*. On the available evidence, horizontal resistance is by far the more important. Vertical resistance associated with hypersensitivity does indeed occur, but in the survey by Hooker and le Roux (1957) it was absent from most maize strains and unimportant in all but one of the few strains in which it was present. Of 85 maize strains from Iowa and Wisconsin tested at Wisconsin, 16 had high resistance in the field. Of these 16 only 1 was hypersensitive, and that to only 2 out of the 15 isolates of *P. sorghi* used. In Iowa, 160 strains of maize had high resistance in the field; of these only 35 were hypersensitive, and that only to some of the isolates of *P. sorghi* but not to others. If without absolute proof we assume this stable field resistance unassociated with hypersensitivity to be horizontal resistance, then horizontal resistance has played the major role in protecting maize against *P. sorghi*. How great this protection can be is forcefully brought home when one sees occasional highly susceptible inbred lines blasted by rust to show what damage the pathogen can do when most of the resistance to it is lost.

Tropical rust of maize is caused by *Puccinia polysora*. *Puccinia polysora* was first described on grasses. So little damage does it do to maize in the Western Hemisphere that it was not identified on maize until 1941 (Cummins 1941), although examinations of old herbarium specimens of maize showed that it had been widely collected in the past. It was collected as far north as Massachusetts almost a century ago, but most collections were from tropical America and the southern United States. It suddenly became important when it crossed the Atlantic Ocean in 1949 or shortly before. In Africa it found maize with its resistance dissipated after more than four centuries out of contact with the pathogen, and tropical rust burst out in a great epidemic. Maize strains were killed before they could ripen, and the epidemic swept across tropical Africa and into the Indian Ocean as far away as the islands of Mauritius and Réunion. By 1954 the epidemic had begun to abate, largely because farmers grew the most resistant varieties they could find (Cammack 1961); and finally enough resistance was accumulated to make tropical rust in Africa a relatively unimportant disease.

While the epidemic was at its height in Africa, maize varieties were introduced and tested for performance in the field. Varieties were classified as susceptible if they were severely damaged, and resistant if they were not. Stanton and Cammack (1953) published the findings which are summarized in Table 7.6. All the lines from tropical America had considerable resistance; all the lines from Africa and Asia were susceptible. Asia, like Africa, had been free from *P. polysora;* and Asian maize, like African maize, lacked resistance. Maize from the United States was intermediate in resistance. Because African and Asian maize originated in America, we have here evidence of how resistance was dissipated over the centuries when maize and *P. polysora* were separated in Africa and Asia.

To judge by the maize of tropical America, most of the resistance is horizontal. Little is vertical resistance associated with hypersensitivity. Storey *et al.* (1958) tested nearly 200 strains of maize: 6 from Mexico,

TABLE 7.6

Resistance to Tropical Rust of Maize Varieties from Several Sources [a]

Source	No. of varieties	Resistance
West Africa	31	None
East Africa	7	None
South Africa	2	None
India	18	None
Ceylon	7	None
Malaya	2	None
United States	20	Variable
Mexico	22	Considerable
Caribbean	3	Considerable
Venezuela	2	Considerable

[a] Data of Stanton and Cammack (1953).

19 from Columbia, 118 from Trinidad, and 51 from Costa Rica. They were tested in the greenhouse for resistance associated with hypersensitivity in juvenile plants. Most strains gave no hypersensitive reaction at all. Of the few that did show hypersensitivity none was homozygous for it.

Further evidence that resistance to *P. polysora* is mainly horizontal comes from an experiment in west Africa. Twelve different maize varieties were grown in assay plots in almost all the recognized vegetation zones of Nigeria, Southern Cameroons, and Ghana. The varieties stayed constant in resistance relative to one another, i.e., there was no evidence of a differential interaction with rust in the different localities.

To summarize, horizontal resistance in maize to both *P. sorghi* and *P. polysory* seems to be not only the common form of resistance but also a highly efficient form. In America it has controlled both these pathogens well. In Africa the evidence is strong that the reaccumulation of dissipated horizontal resistance now holds tropical maize rust strongly in check.

Evidence about potato blight is easier to interpret because in the absence of *R* genes resistance in the potato seems to be mainly or wholly horizontal (see Section 7.13). Substantial horizontal resistance occurs, especially in the late maturing potato varieties more commonly grown in Europe than North Amerca. Within a maturity class great varietal differences exist. The very great difference in horizontal resistance between, say, the susceptible variety Up-to-Date and the resistant Pimpernel, both in much the same maturity class, illustrates this.

The catastrophic epidemics of potato blight in Europe in the 1840's were due primarily to poor resistance and not to weather especially favorable to the disease. Bourke (1964) analyzed data about the weather in Ireland at the time, and concluded that nowadays such weather would not cause a severe epidemic. Relic floras of potatoes originating from Europe in preblight days are mostly intensely susceptible to blight (Van der Plank, 1968). Clearly, resistance has increased substantially since the 1840's. This increase has been paralleled in experiments by Simmonds and Malcolmson (1967). They took potato varieties from the Andes, the ancestral home of *Solanum tuberosum*. These varieties have no known *R* genes. Blight is unimportant on the Andes, and the Andean varieties, like the European varieties before 1845, are mostly very susceptible. Nevertheless, it was possible to raise the level of resistance of the Andean varieties to that of modern European varieties simply by mass selection during three sexual generations.

Maize recovered resistance to tropical rust in Africa. The recovery was rapid. Potatoes gained resistance to blight in Europe. The gain was reasonably rapid, if one remembers that potatoes are propagated by clones. The rapidity suggests that most variation in horizontal resistance is controlled by relatively few genes. Also, maize in Africa and potatoes in Europe are not lacking in vigor. Increased horizontal resistance was obtained without a crippling loss of fitness. This stresses the bright side of horizontal resistance.

The other side is less bright. It seems unquestionable that, at least in outbreeding plants or plants propagated by clones, increased horizontal resistance reduces the fitness of the host plants when the pathogen is absent, even though the reduction may not be severe or crippling. See Section 5.6. The evidence of tropical rust in Africa or potato blight in Europe can only mean that the fittest host type to survive in the absence of disease differs from the fittest type when the pathogen is present and conditions favor disease. That, too, is the evidence of the Vertifolia effect, i.e., the loss of horizontal resistance when breeders select cultivars protected from disease by vertical resistance (Van der Plank, 1968). That, too, is the evidence of the long-known fact that epidemics are often at their most catastrophic when host and pathogen meet for the first time, and thereafter abate without the intervention of fungicides or a change of climate or other environmental factor. Powdery mildew of grape vines is an example that closely resembles potato late blight. *Uncinula necator* reached western Europe at about the same time as *Phytophthora infestans,* caused similar destruction, and then abated in a similar way. Catastrophic initial epidemics followed by abatement without apparent external cause should always make one suspect that excess horizontal resistance has disadvantages.

In Section 5.6 the disadvantage of excess horizontal resistance to endemic fungus disease was stressed. With endemic disease excess resistance can be defined as more resistance than is needed to keep disease at a level where it does little harm to the host plant, or less harm than more resistance in the host plant would do. With endemic disease a disadvantage could accumulate over hundreds of years, and the effect of excess resistance is likely to be clearer and more important than it is with diseases of annual crops. For this reason we believe that the study of endemicity is perhaps the best path to an understanding of horizontal resistance, its possibilities, its advantages, and its disadvantages.

If horizontal resistance increases with an accumulation of homozygotes as a form of gene duplication, and if loss of heterosis reduces vigor, then useful horizontal resistance is likely to be attained only if not many genes are greatly involved. We have inferred this to be the case with tropical maize rust and potato blight. Considered more generally, the inequality of effect of resistance genes is a factor likely to determine how far horizontal resistance can usefully be included in a breeding program for clonal crop plants or outbreeders.

About two categories of disease information on the possibility of greatly increasing horizontal resistance is scarce. These are diseases of plants like wheat that are essentially inbreeders, and diseases caused by pathogens other than obligate parasites or near obligate parasites like *Phytophthora infestans.* We are also ignorant of the extent to which horizontal resistance already occurs. But for the epidemic in Africa, *Puccinia polysora* would have been written off as a minor parasite incapable of harming maize greatly. We now see it to be potentially as destructive in the tropics as any of the worst pathogens, but controlled by a high dose of horizontal resistance. Many, perhaps most, of the recorded plant diseases have little economic importance. How many are controlled by resistance rather than by environment? We cannot answer this question because we do not as a rule pay much attention to diseases of little economic importance. We concentrate on the destructive diseases, which are the diseases against which resistance is inadequate; and by this concentration we get a biased assessment, and underestimate the prevalence of resistance. We should try to know more of unimportant diseases and endemic disease.

7.19 IMMUNITY AS EXTREME VERTICAL RESISTANCE

Immunity, as we follow convention and use the word here, is absolute. Resistance can be partial and occur in degrees; plants may be
weakly resistant or strongly resistant. There are no grades of immunity;
a plant is immune—a nonhost to use current jargon—or not immune.
Thus, wheat is immune from *Phytophthora infestans.* Because to any
given fungus, bacterium, virus, or nematode there are far more nonhosts than hosts, immunity is the rule and susceptibility the exception.

Immunity resembles vertical resistance closely in two of its features,
and we classify it as vertical resistance in its extreme form.

First, immunity involves a differential interaction between genotypes
of the host and genotypes of the pathogen. *Puccinia graminis* attacks
grasses but not sunflowers; sunflowers are immune. *Puccinia helianthi*
attacks sunflowers but not grasses; grasses are immune. Differential interaction is involved here just as much as when Canadian race *C*22 of
Puccinia graminis tritici attacks Sr_6-types of wheat but not Sr_{9d}-types,
and race *C*10 attacks Sr_{9d}-types but not Sr_6-types. The difference between vertical resistance and immunity is largely one of taxonomic status. In vertical resistance the taxa involved are usually races of the
pathogen and varieties or species of the host plant. In immunity the
taxa are usually species, genera, or higher taxa of both the pathogen
and the host plant. These greater taxonomic differences in turn reflect
the greater chemical and genetic differences that in immunity defeat the
pathogen.

Second, immunity and vertical resistance sometimes resemble each
other in hypersensitivity symptoms and in phytoalexins they produce.
Thus, Müller (1950) used *Phytophthora infestans* to inoculate *Phaseolus, Brassica,* and seven other immune angiosperms not members of the
Solanaceae, and obtained hypersensitivity reactions resembling those
produced in vertically resistant potatoes. In phytoalexin studies, too, little or no distinction can be made between incompatibility resulting from
vertical resistance and that resulting from immunity. (For a discussion
about the association of phytoalexins and hypersensitivity with vertical
resistance and immunity, see Section 7.23. The association is indirect
and peripheral.)

There are two entirely separate problems of immunity to be discussed. The first is concerned with the mechanism of immunity: how
avirulence leads to the pathogen being excluded or destroyed. This is not
strictly an immunity problem, because avirulence reactions may well be

the same in both immunity and vertical resistance. Gene identity is involved, and discussion links up with the first (Flor's) gene-for-gene hypothesis.

The second problem is why in immunity the pathogen cannot become virulent: Why, for example, *Puccinia graminis* cannot or does not make the leap to become virulent on sunflowers. This is the true immunity problem, in that it distinguishes immunity from mere vertical resistance. It involves gene quality, and the discussion, in Section 7.24, is based on the second gene-for-gene hypothesis.

7.20 ANTIGENS OF HOST AND PARASITE IN VERTICAL RESISTANCE AND IMMUNITY

Flor established his gene-for-gene hypothesis on the relation between resistance in flax and virulence in the flax rust fungus *Melampsora lini*. Flor and co-workers (Doubly *et al.,* 1960) followed the hypothesis up with a serological study of host–pathogen relations. They used four flax varieties of varying resistance and four isolates of *M. lini* of varying virulence. As antigens they used globular protein fractions from the flax varieties and from uredospores of the rust isolates. They prepared antisera in rabbits. As is shown in Table 7.7, the host–pathogen interaction was susceptible when titers of rust antiserum against flax antigens were relatively high (1:160 or 1:320) and resistant when the titers were low

TABLE 7.7

Titers of Rust Antiserum against Flax Antigens in Relation to the Resistance of Flax against *Melampsora lini* [a]

Flax variety	Race of rust			
	1	210	19	22
Bison	S	S	S	S
	1:320	1:320	1:320	1:160
Koto × Bison 7	R	S	S	S
	1:40	1:320	1:320	1:320
Cass × Bison 7	R	R	S	S
	1:40	1:20	1:320	1:160
Ottawa 770 B × Bison 7	R	R	R	S
	1:40	1:20	1:40	1:320

[a] Data of Doubly *et al.* (1960). Here *R* stands for resistant, and *S* for susceptible.

(1:20 or 1:40). Successful parasitism depended on a substantial quantity of antigens being shared by host and pathogen; and greater disparity between antigens of host and parasite led to vertical resistance. Evidently, when a resistance gene is introduced into the host (in the present instance by backcrossing on the flax variety Bison), it changes the appropriate antigen in the host and creates disparity until such time as the pathogen in turn mutates to virulence and reduces the disparity again.

Dealing with an example of immunity instead of vertical resistance, Wimalajeewa and De Vay (1971) compared the antigens of *Ustilago maydis* with those of susceptible maize and immune barley. *Ustilago maydis* and maize had antigens in common; *U. maydis* and barley had no antigens in common. Oats were intermediate. A diploid line of *U. maydis* attacked 3-day-old oat seedlings but not 6-week-old seedlings; and in conformity with this, the diploid line of *U. maydis* and 3-day-old oat seedlings shared antigens, while the antigenic sharing was much weaker between *U. maydis* and 6-week-old seedlings. Wimalajeewa and De Vay showed that in both maize and *U. maydis* the shared antigens were proteins loosely held in the ribosomes.

Wimalajeewa and De Vay reviewed the literature, including the literature of animal host–parasite relations. There is a close connection between susceptibility and the sharing of antigens by host and parasite, and between resistance and a disparity in the antigens of host and parasite.

The genetics of this is not altogether clear. Antigens common to host and parasite would imply genetic material common to host and parasite; and antigenic disparity would imply genetic disparity. But we must beware of oversimplification. When there are too many genes for virulence, i.e., when there are genes for virulence in the pathogen without the matching resistance genes in the host, the genetic disparity between host and pathogen is just as great as when there are too few genes for virulence, i.e. when there are resistance genes in the host without the matching virulence genes in the pathogen. This is not reflected antigenically. There is antigenic disparity when there are too few virulence genes but not when there are too many. (In Table 7.7, too few virulence genes are shown by an *R* reaction. There are too many virulence genes in, e.g., the combination of Bison flax with races 19, 22, or 210 of *M. lini*.)

Here is a new theory of why *M. lini's* virulence is recessive: Excess virulence is antigenically neutral. The previous paragraph's analysis shows this. If virulence were antigenically active, and antigenic disparity accompanied resistance even secondarily, too many virulence genes would be as harmful to the pathogen as too few.

Although most work has been done with fungus pathogens, Schnathorst and De Vay (1963) working with *Xanthomonas malvacearum* found that resistant varieties of cotton shared fewer antigens with the bacterium then susceptible ones. Bacteria and fungi seem to be in line.

7.21 IMMUNITY AGAINST VIRUSES

Immunity patterns against viruses resemble those against fungi and bacteria. There is the same differential interaction between host and pathogen. In the genetic control of incompatibility the pathogen is as deeply involved as the host.

The preceding section was given over to the proposition that antigenic proteins hold the key to the compatibility or incompatibility of host and parasite. The evidence from some viruses makes this proposition questionable. This evidence suggests that the compatibility or incompatibility of the proteins may be no more than the secondary consequence of a primary compatibility or incompatibility at nucleic acid level. That is, in immunity the clash between host and parasite may primarily be between nucleic acids, with proteins the reflections. The relevant point is that several viruses have no protein yet are subject to immunity control.

Tomato bunchy top, potato spindle tuber, and citrus exocortis viruses, discussed in Section 1.7, appear to contain no protein. All the evidence now available agrees that they are naked particles of nucleic acid without the protein coat that most other viruses have. The unmixed long particles of tobacco rattle virus studied by Harrison and Woods (1966), Sänger (1968a,b), and Huttinga (1972) also appear to be free from protein. Potato spindle tuber virus has been the subject of much research recently. Rubbed on to the leaves of *Scopolia sinensis* this virus produces small necrotic spots (Singh, 1971). This reaction makes *S. sinensis* a useful local-lesion indicator plant for the virus. But our concern here is with host–virus incompatibility. No virus protein is involved, because there is no virus protein. True, the virus may in some way combine with host protein and thus escape destruction by the host's cellular ribonucleases. But it is improbable that there would be incompatibility between the host and its own proteins; and there seems to be no way of avoiding the conclusion that the virus nucleic acid itself is involved in the incompatibility. (We write here of incompatibility instead of immunity. The initial reaction of potato spindle tuber virus is one of local lesions and hypersensitivity; but the virus ultimately escapes to cause a systemic veinal necrosis. The grip of the host's defences is not complete. This may conform with what is said in the final paragraph of this section.)

Judged by the very scant evidence now available, the incompatibility between *S. sinensis* and potato spindle tuber virus is much the same as the incompatibility between, say, flax and an avirulent isolate of *M. lini*. Unless some fundamental difference between the two incompatible combinations can be discovered, one may have to scrap the protein antigen theory of resistance to flax rust, and dig deeper. The coatless viruses need more study. Just as genetics has gained immeasurably from the study of bacterial viruses, so the science of immunity, by far the most important in all host–parasite relations, can gain by stripping away as many complications as possible and concentrating on the essentials for the existance of the least complicated parasites yet known.

· The plant host seems to have more difficulty becoming immune to viruses than to fungi. If, to compare fairly, one compares only obligate parasites, one finds that there are no obligately parasitic fungi with host ranges nearly as wide as those of viruses which attack many diverse families of monocotyledons and dicotyledons, or which attack angiosperms and fungi or angiosperms and nematodes or angiosperms and insects. It would not be difficult to imagine a model in which the host plant's difficulty in controlling pathogens by immunity increases as the complexity of the pathogen's genome decreases.

7.22   THEORY OF IMMUNITY OR VERTICAL RESISTANCE THROUGH
PHYTOALEXINS OR HYPERSENSITIVITY A GENETIC MISFIT

Phytoalexins are chemical inhibitors synthesized in the host plant. They are potent fungicides and antibiotics generally. The phytoalexin theory of resistance is that invasion of the host by the pathogen induces the host to produce phytoalexins that destroy the pathogen.

There is now a large literature of phytoalexins. Two facts seem to be established beyond dispute. First, phytoalexins are commonly associated with vertical resistance or immunity; that is, they are associated with differential interactions between host and pathogen, in which the part of the pathogen is highly specific. Second, phytoalexins are characteristic

of the host alone—peas produce pisatin, beans phaseollin, sweet potatoes ipomeamarone—and the pathogen, so far from being a specific
part of the process, can be replaced even by inorganic chemicals such
as copper chloride. These two facts seem irreconcilable in any theory of
phytoalexins controlling vertical resistance. The phytoalexin theory of
vertical resistance or immunity has been a misfit from the start.

In long overdue experiments Király *et al.* (1972) have questioned
the phytoalexin theory of resistance. They showed phytoalexins to be
the effect, not the cause, of the pathogen's death. They used the potato
and *Phytophthora infestans,* and determined the occurrence of hypersensitivity and the potato's phytoalexin, rishitin. Because hypersensitivity and the phytoalexin were produced concomitantly, we can consider them together. They also used wheat and *Puccinia graminis,* and
beans and *Uromyces phaseoli.* They induced a hypersensitive type of
necrosis in tissues of potato tubers, wheat leaves, and bean leaves infected with compatible (virulent) isolates of the pathogens, when these
pathogens were inhibited or prevented from further growth in host tissues. The hypersensitivity thus induced, and the phytoalexin that accompanied it, did not differ from what was found when the tissue was
infected with incompatible (avirulent) isolates. The chemical inhibitors
and the compatible isolates separately did not induce a hypersensitive
type of necrosis or phytoalexin production. Sonicated or chloroform-
treated mycelium of compatible isolates caused hypersensitivity and phytoalexin production just as incompatible isolates did.

From these and other experiments Király *et al.* concluded that resistance first requires the killing or inhibition of the fungus; and that
this killing or inhibition then causes endotoxins to be released from the
pathogen which make the host react hypersensitively and produce phytoalexins. This makes genetic sense. The sequence of events set off by
the differential interaction between host and pathogen continues so far
as the killing or inhibition of the pathogen, so that vertical resistance or
immunity preserve the differential interaction. Here the original sequence discontinues; and after the discontinuity a new sequence begins
in which host–pathogen interactions are defaced. Two difference genetic
systems are involved. Up to the discontinuity the host's genes for resistance or susceptibility and the pathogen's genes for avirulence or virulence are active. After the discontinuity the pathogen's genes are elimi-
nated by the death of the pathogen; and there is no evidence to
implicate the vertical resistance or susceptibility genes of the host, because copper chloride can replace the pathogen.

7.23   AN INTERLUDE ON RESISTANCE AGAINST SECONDARY
INFECTION. PREFORMED LOCALIZED RESISTANCE.
A ROLE FOR PHYTOALEXINS
AND HYPERSENSITIVITY

Phytoalexin production follows the death or inactivation of the parasite. At the concentrations at which they are produced, phytoalexins destroy many plant pathogens. Phytoalexins are produced widely, in angiosperms at least. Adding this together, we see phytoalexins and
hypersensitivity as protecting the host against secondary parasites enter-
ing through breaches left open in the host's defenses by the death or inactivation of the primary parasite.

With the distinction between primary and secondary infection in
mind, we can reassess the experiments of Müller and Börger (1941).
Part of the cut surface of tubers of a potato variety with an R gene was
inoculated with zoospores of an avirulent isolate of *Phytophthora infestans*. After some time the whole cut surface was inoculated with zoospores of a virulent isolate. The virulent isolate grew freely on the surface not previously inoculated with the avirulent isolate but poorly on
the part of the surface previously inoculated. Moreover, they found the
previously inoculated tissue to be resistant to infection not only by *P.
infestans* but by *Fusarium caeruleum* as well.

These experiments of Müller and Börger started the phytoalexin
theory. Results ought to have been taken at face value. What they
showed was that a pretreated surface was resistant to secondary infection. There was good evidence in their experiments for a process acting
against secondary infection, but no real evidence that this same process
acted against primary infection. So too in the many experiments of others that followed Müller and Börger the evidence, when critically evaluated, is found to be real evidence only for a process that could act
against secondary invasion. The conclusion jumped to by Müller and
Börger and their followers, that hypersensitivity and phytoalexins are
the basis of resistance to primary infection, can be seen as a huge non
sequitur.

There is a consensus that a process that occurs commonly in a population is likely to be a process useful to that population. Hypersensitivity and phytoalexin production are common, perhaps universal, processes in angiosperms. (Other phyla have still to be examined closely.)
They are processes of the host alone; typically they occur in a parasitic
no man's land, between the death or inactivation of the primary para-

site and before invasion by a secondary parasite. Adding this together, we infer that hypersensitivity and phytoalexin production are processes highly useful to the host population.

On our hypothesis, then, resistance to secondary invasion is highly useful to the host population. Secondary infection does not play an important part in plant pathology. How can resistance to secondary infection be very important, if secondary infection itself is not very important? One must not invert the reasoning. Secondary infection may seem relatively unimportant; but that is simply because hypersensitivity and phytoalexin production keep it relatively unimportant. Its relative unimportance testifies to the efficiency of the resistance against it. Secondary infection is not inherently a minor threat; it is a threat adequately met and well contained.

The plant surface is crowded with microorganisms, many of them potentially pathogenic. What to us may seem to be no more than a barely visible fleck is to them a wide breach through which they can invade. The host plant, we suggest, closes this breach quickly with a chemical barrier, until it has time to build the physical barriers that go with the healing of wounds.

Reference to primary and secondary pathogens, or invasions, does not imply that two different species of pathogen are necessarily involved. Primary and secondary pathogens may or may not differ in species. In Müller and Börger's experiments, for example, the species were the same when *P. infestans* followed *P. infestans* but differed when F. *caeruleum* followed *P. infestans*.

The theory that phytoalexins are preformed antibiotics protecting the plant against secondary infection could explain why plants sometimes form phytoalexins after infection by bacteria (Stholasuta *et al.,* 1971) and viruses (Bailey and Ingham, 1971), although there is yet no evidence that phytoalexins are active against bacteria and viruses. It seems that as far as secondary invaders are concerned it is only the fungi that the host plant uses phytoalexins against. Viruses cannot penetrate dead tissue, and hypersensitivity needs no reinforcement. Many fungi could use dead tissue as a portal of entry, and hypersensitivity needs phytoalexins as reinforcement against secondary invasion by them. Bacteria, unlike fungi, are incapable of aggressive assault; and one infers that continuous dead tissue is in itself a substantial barrier. The comparative evidence from the three great groups of pathogens, summarized, is that phytoalexins are a common consequence of primary infection by incompatible fungi, bacteria, and viruses, and a common protection against secondary infection by some fungi; they seemingly play no part in the host plant's defenses against bacteria and viruses. Events have been odd. It is accepted in the literature that the phytoalexin concept traces back to Müller and Börger's (1941) experiments with *Phytophthora infestans* in potato tubers. But *P. infestans* grows poorly, if it grows at all, on dead potato tissue; and it is unlikely that phytoalexins, as distinct from hypersensitivity, are needed by the potato against blight.

Resistance to secondary infection often follows vertical resistance, presumably because the processes of vertical resistance leave breaches in the host plant's defenses. It is not strongly associated with horizontal resistance. Horizontal resistance seldom kills the pathogen as it enters. As a rule, it neither provides the host plant with a strong trigger to hypersensitivity and phytoalexin production, nor, it seems, does it leave an open breach for secondary invaders to enter through. Horizontal resistance can indeed cause necrosis, but usually does so only late in the lesion's development and not at the boundary of the pathogen's advance into host tissue. The greater development of necrosis at the center of an old potato blight lesion in a horizontally resistant variety is an example.

The association between hypersensitivity and vertical, but not horizontal, resistance makes a convenient tag by which vertical resistance can be recognized. For purposes of tagging it is immaterial that the association is peripheral and not causal.

The theory of resistance to secondary invasion can be extended to a general theory of preformed, localized resistance by phytoalexins and hypersensitivity. Primary parasites are not alone in inducing the production of phytoalexins and hypersensitivity. Chemicals and wounding can do so as well, with varying efficiency, and can thus replace primary parasites as triggers of resistance.

Resistance is preformed. It is governed genetically by the host alone. It is preformed in that the secondary parasite or parasite following wounding is resisted by phytoalexins and hypersensitivity it had no part in producing, even as a trigger. In being preformed, phytoalexins do not differ from the fungitoxic phenolic compounds found by Walker and Link (1935) in colored scales of onion bulbs, or from the fungistatic phenolic acids found by Martin *et al.* (1957) in the waxy covering of apple and other leaves, or from the phenolic compounds found by Clauss (1961) in the seed coat of pea varieties resistant to foot rot, or from the antifungal compound found by Lampard and Carter (1973) in the cuticular wax of coffee or from the decay-resisting substances in the heartwood of trees (Scheffer and Cowling, 1966) or from avenacin in oats, hordatins in barley, and the few other known preformed anti-fungal compounds.

Resistance is localized. It is concentrated at or near the breach in the host plant's defenses. In this localization, phytoalexins and hypersensitivity differ sharply from the diffused preformed substances in colored onion bulb scales, apple cuticular waxes, pea seed coats, or coffee cuticular wax. This difference is probably vital. Resistance must curb or destroy the pathogen without destroying the host plant. To develop resistance by means of preformed diffused chemical substances, the host plant has either to produce substances selectively harmful to pathogens but not to itself, or to banish these substances to bulb scales, cuticular waxes, seed coats or other zones of little or no metabolic activity. But the host plant's difficulty is resolved if resistance is localized instead of diffused. A few cells of the host plant can be expended without much harm to the plant; and with localization a workable resistance is possible with substances or processes indiscriminately active against both host tissues and pathogen.

To summarize, we see phytoalexins and hypersensitivity as providing preformed resistance, which is resistance present before invasion by the pathogen it resists. This is directly opposed to the theory of phytoalexins and hypersensitivity as responses to the pathogen they must provide resistance against. Localization removes the great limitation on preformed resistance. Resistance to secondary infection is a special case of preformed, localized resistance.

7.24   Immunity Compared with Vertical Resistance

The counterpart in the pathogen of immunity or vertical resistance in the host is avirulence. In relation to avirulence there is no known difference between immunity and vertical resistance. Sections 7.19, 7.20, and 7.22 were about avirulence phenenoma that are common to both immunity and vertical resistance: the disparity between antigens of host and pathogen, and the production of phytoalexins and hypersensitivity as peripheral end products. There seems to be no way in which an avirulent pathogen, in the absence of its virulent mutants, could be used to distinguish between immunity and vertical resistance.

It is virulence in the pathogen that distinguishes vertical resistance from immunity in the host. In vertical resistance, virulence in the pathogen exists. In immunity, it does not. This is a matter of definition. If for convenience one thinks of virulence as a mutant of avirulence, then in vertical resistance the mutation can and does take place. In immunity the mutation to virulence is forbidden; appropriate mutation is either destructive to the phenotype or beyond the ability of the genotype of the pathogen. On either alternative one can understand why immunity is usually associated with higher taxa in both host and pathogen than is vertical resistance (see Section 7.19).

We have already seen, in Section 7.11, how wide is the range between virulence and avirulence in their effect on fitness to survive. Virulence in *Phytophthora infestans* on resistance gene R_4 is as fit to survive as avirulence. But combined virulence in *Puccinia graminis* on Sr_6 and Sr_{9d} was conspicuously unfit in Canada during the decades covered by Fig. 7.1. It is easy to visualize that an extension of the range could extend the unfitness of virulence to the point where virulence becomes forbidden.

As part of the discussion about the second gene-for-gene hypothesis it was postulated in Section 7.8 that the gene for avirulence has a useful primary function in the pathogen. In this primary function virulence, it was further postulated, can substitute for avirulence, sometimes well, sometimes badly.

Mutation from avirulence to virulence on gene R_4 is reversible; and it was suggested in Section 7.7 that this mutation involved only an unimportant substitution of one amino acid side chain for another. But with other diseases and other genes, substitution could more drastically widen the difference between avirulence and virulence. More genes could be involved, and, with this, changes in more amino acid side chains. The effect of substitution could also be increased by a greater change in the side chain and a correspondingly greater change in the configuration of the coded protein. Nor is there reason to assume that only reversible point mutations are involved. Irreversible chromosomal mutations could make virulence differ so much from avirulence that it would not function adequately in its primary useful role, and the phenotype would thus cease to be viable. As soon as the virulent phenotype ceased to be viable, the border would be crossed from vertical resistance to immunity.

Alternatively, the general proposition that large mutations are rarer than small mutations, i.e., that the larger the mutation the less often it is likely to occur, might be involved. If large mutations were needed from avirulence to virulence on several immunity genes in the host, and if each mutation caused much loss of fitness, the pathogen might not be able to make them simultaneously.

7.25 VIRULENCE AND AGGRESSIVENESS

Virulence and avirulence in the pathogen are the counterparts of vertical susceptibility and resistance in the host. Aggressiveness and un-aggressiveness in the pathogen are the counterparts of horizontal susceptibility and resistance in the host.

For disease to occur, the pathogen must be virulent. It may be virulent and strongly aggressive or virulent and unaggressive (weakly aggressive).

The concepts of virulence and aggressiveness are entirely distinct. Virulence involves gene diversity, probably largely through mutation. Aggressiveness may well involve enzyme dose (as distinct from enzyme diversity) and the switching on and off of enzyme action. Great aggressiveness probably does not involve any processes not also present in less aggressive isolates; and mutation if it occurs probably has no pleiotropic effects that substantially alter the kind of process.

7.26 SHARED PHENOTYPIC EFFECTS OF THREE GENOTYPES

Three genotypes affect the infection rate. The rate is reduced if horizontal resistance in the host is increased, if aggressiveness in the pathogen is decreased, or if unnecessary virulence in the pathogen replaces avirulence matching a strong resistance gene, as defined in Section 7.9. All three affect the infection rate by affecting the basic infection rate R, the period of infectiousness i, or the period of latency p. There is nothing specially noteworthy about this; these are the only ways in which the rate can be affected.

The effect of increased horizontal resistance or decreased aggressiveness in decreasing the infection rate is fairly obvious, and has been discussed earlier. The effect of changing from avirulence to unnecessary virulence is quantitatively assessed as the relative half-life of the virulent pathogen compared with that of the avirulent pathogen. In turn, the relative half-life can be quantitatively interpreted as a change of infection rate by the appropriate equation [Equation (4.1) in Van der Plank, 1968]. The relative half-life and associated change in the infection rate measure the strength of the vertical resistance gene.

7.27   SOME COMMENTS ABOUT TERMINOLOGY

The terms vertical resistance and horizontal resistance are geometric variants of differential resistance and nondifferential resistance. Thus, horizontal resistance is resistance that does not vary with variations of pathogenic aggressiveness; resistance plotted against races or isolates of the pathogen would give a straight horizontal line. The analytic variants, differential resistance and nondifferential resistance, are clear, but involve a clumsy "non" word. The alternatives, differential resistance, and uniform resistance, carry a connotation that could mislead. For example, some potato cultivars without R genes have much resistance to blight in the foliage but little in the tubers; to call this resistance uniform might well mislead.

The term field resistance has nothing to commend it and much to condemn it. It should be scrapped, for three reasons. First, it is used oppositely for different diseases; it is used to mean horizontal resistance to potato blight, but vertical resistance to fusarium wilt of tomatoes. Second, all resistance is resistance in the field; what is field resistance supposed to distinguish? Third, the term has been abused and distorted beyond recognition. For example, Killick and Malcolmson (1973) used the term field resistance for phenomena observed exclusively in the laboratory and mimicking only vaguely and remotely what goes on in the field.

The term, partial resistance, sometimes used for horizontal resistance, is tautological and unacceptable. On current definitions, discussed in Section 7.1, resistance as opposed to immunity is always partial, irrespective of whether it is horizontal or vertical; and the term cannot be used to describe one form of resistance to the exclusion of another.

Another term with nothing to commend it is true resistance, used by Müller (1953) and Kiyosawa (1971) to mean vertical resistance. The implication that horizontal resistance is false does not bear scrutiny.

The term race specific resistance or, simply, specific resistance for vertical resistance is inappropriate because differential interactions between host varieties and pathogenic races can occur without race specificity. That is, all the relevant races may attack all the relevant varieties, but not uniformly. Christensen and Graham's (1934) experiments with *Helminthosporium gramineum* on barley are an example. The term is all very well when host plants can be classified as resistant or susceptible with nothing in between, but is inapplicable when gradations occur in resistance and differential interaction is involved. It is a mistake to couple a definition to a preconception that intermediates in resistance do not occur.

Race is the standard word in the population sciences, including population genetics, for a distinct population within a species or subspecies. There have been attempts in plant pathology to replace it by the word pathotype. They should be resisted. The issue is not whether race is a good word or a bad word. Race is the used word; that is what matters. Few things are likely to be more destructive to a science than a cosy jargon of its own that shelters it from the impact of ideas from other sciences. The genetics of host–pathogen relations must develop under illumination from population genetics generally, and a specialized terminology that leaves it in shadow will inevitably retard progress. In saying this we do not for a moment concede that the used word, race, is badly chosen and needs to be apologized for. Nor are we aware of any wide move to replace it in the population sciences generally, or of any confusion of meaning that would make such a move desirable.

There are two circumstances in which it is useful to add an adjective and describe a race as a pathogenic race. One may need to distinguish pathogenic races from geographical races, climatic races, and the like. The need for this does not occur in this book. Or one may need to emphasize that one is referring to the pathogen and not the host, as when one writes of interactions between pathogenic races and host varieties. But ordinarily the context makes the adjective, pathogenic, superfluous. If an adjective is to be used, the term should be pathogenic race—or race of the pathogen—and not physiological race, which is no more than a tedious addiction. One may assume that all races, be they vertical races or horizontal races, pathogenic races or geographical races, have a physiology.

The concept of race can easily be overworked. Virulence and the frequency of virulence in the population are commonly more appropriate for study. In Sections 7.5 and 7.6 we studied the frequency of virulence in populations, and the numerical data in Tables 7.2 and 7.3 refer to virulences. Races are not mentioned in these tables (except in a footnote referring to a race mentioned in the literature). Too commonly in the literature, races are discussed when it would be more accurate and apt to discuss virulence. For example, if an isolate of *Phytophthora infestans* is found to attack a potato with the gene R_1, what is shown is that it is virulent on R_1. What is often written is that it is of race 1. This is incorrect because, even if one allows for no more than ten R

genes, it could belong to any one of 512 races all virulent on R_1. If a new race must be created for every change of pathogenicity, then we must envisage more than a million races of *Puccinia graminis tritici,* even if we allow for only two types of reaction, susceptible and resistant, without intermediates and without recognition that each of these races could with equal logic be further divided to reflect countless variations of aggressiveness. We doubt whether the concept of race as a taxon within a species or subspecies was ever meant to record the vagaries of the occurrence or duplication of single genes.

In presenting numerical data the sooner we switch from races to virulence frequencies, the quicker the population genetics of host–pathogen combinations will develop. Granted, it is still impossible with many species of pathogen to determine the virulence of an isolate on a particular resistance gene; the appropriate isogenic (or near isogenic) host lines are not yet available. We must then devise an artificial system of races, but should recognize them to be covers for ignorance. Granted, too, that even if the appropriate isogenic lines are used, it is a form of shorthand to try to present results in terms of concocted races. But this shorthand would be unnecessary if only authors would state what resistance genes the isolates of the pathogen were tested against; a straight statement of an isolate's virulences would then be as short as any statement on races, just as informative, and easier to interpret.

When numerical data are not being discussed, the word race can sometimes be used as a convenient abstraction. But even here it would be well to preserve distinctions. In describing experiments authors often write that they inoculated plants with a race virulent on some particular resistance gene. Here the word isolate or culture is better. A race is a population, with all the variation that goes with a population, and this meaning is not intended. We have already (in Section 7.12) protested against confusing isolates with races; and more precision in terminology would not be amiss.

Bibliography

Abrahamson, S., Bender, M. A., Conger, A. R., and Wolff, S. (1973). Uniformity of radiation-induced mutation rates among different species. *Nature (London)* **245**, 460–462.

Albersheim, P., and Anderson, A. J. (1971). Host-pathogen interactions. III. Proteins from plant cell walls inhibit polygalacturonases secreted by plant pathogens. *Proc. Nat. Acad. Sci. U.S.* **68**, 1815–1819.

Albersheim, P., Jones, T. M., and English, P. D. (1969). Biochemistry of the cell wall in relation to infective processes. *Annu. Rev. Phytopathol.* **7**, 171–194.

Allen, P. J. (1965). Metabolic aspects of spore germination in fungi. *Annu. Rev. Phytopathol.* **3**, 313–342.

Anderson, A. J., and Albersheim, P. (1972). Host-pathogen interactions. V. Comparisons of the abilities of proteins isolated from three varieties of *Phaseolus vulgaris* to inhibit the endopolygalacturonases secreted by three races of *Colletotrichum lindemuthianum*. *Physiol. Plant Pathol.* **2**, 339–346.

Anonymous (1953). Some further definitions of terms used in plant pathology. *Trans. Brit. Mycol. Soc.* **36**, 267.

Anonymous (1954). Verslag van de enquete over het optreden van de aartappelziekte, *Phytophthora infestans* (Mont.) de Bary, in 1953. *Jaarb. Plantenziektenkundig Dienst Wageningen*, pp. 34–53.

Bailey, J. A., and Ingham, J. L. (1971). Phaseollin accumulation in bean (*Phaseolus vulgaris*) in response to infection by tobacco necrosis virus and the rust *Uromyces appendiculatus*. *Physiol. Plant Pathol.* **1**, 451–456.

Bailey, N. T. J. (1957). "The Mathematical Theory of Epidemics," 194 pp. Griffin, London.

Baker, R., Maurer, C. L., and Maurer, R. A. (1967). Ecology of plant pathogens in soil. VII. Mathematical models and inoculum density. *Phytopathology* **57**, 662–666.

Bateman, D. F., and Millar, R. L. (1966). Pectic enzymes in tissue degradation. *Annu. Rev. Phytopathol.* **4**, 119–146.

Bawden, F. C., and Pirie, N. W. (1959). The infectivity and interaction of nucleic acid preparations from tobacco mosaic virus. *J. Gen. Microbiol.* **21**, 438–456.

Baxter, D. V. (1943). "Pathology in Forest Practice," 618 pp. Wiley, New York.

Bell, A. A., and Daly, J. M. (1962). Assay and partial purification of self-inhibitors of germination from uredospores of the bean rust fungus. *Phytopathology* **52**, 261–266.

Berard, D. F., Kuć, J., and Williams, E. B. (1973). Relationship of genes for resistance to protection by diffusates from incompatible interactions of *Phaseolus vulgaris* with *Colletotrichum lindemuthianum*. *Physiol. Plant Pathol.* **3**, 51–56.

Berkson, J. (1944). Application of the logistic function to bio-assay. *J. Amer. Statist. Ass.* **39**, 357–365.

Berkson, J. (1953). A statistically precise and relatively simple method of estimating the bio-assay with quantal response. *J. Amer. Statist. Ass.* **48**, 565–599.

Bingham, R. T., Hoff, R. J., and McDonald, G. I. (1971). Disease resistance in forest trees. *Annu. Rev. Phytopathol.* **9**, 433–452.

Bliss, D. E. (1946). The relation of soil temperature to the development of Armillaria root rot. *Phytopathology* **36**, 302–318.

Boelema, B. H. (1973). Infectivity titrations of plant pathogenic bacteria. *S. Afr. J. Sci.* **69**, 332–336.

Bonde, R., and Schultz, E. S. (1943). Potato refuse piles as a factor in the dissemination of late blight. *Maine Agr. Exp. Sta. Bull.* No. 416, 230–246.

Bourke, P. M. A. (1964). Emergence of potato blight, 1843–1846. *Nature (London)* **203**, 805–808.

Boyce, J. S. (1938). "Forest Pathology," 600 pp. McGraw-Hill, New York.

Boyle, L. W. (1961). The ecology of *Sclerotium rolfsii,* with emphasis on the role of saprophytic media. *Phytopathology* **51**, 117–119.

Brandes, E. W., and Klaphaak, P. J. (1923). Cultivated and wild hosts of sugarcane or grass mosaic. *J. Agr. Res.* **24**, 247–262.

Branchley, G. H. (1964). Aerial photographs for the study of potato blight epidemics. *World Rev. Pest Contr.* **3**, 68–84.

Brenchley, G. H., and Dadd, C. V. (1962). Potato blight recording by aerial photography. *Nat. Agr. Adv. Serv. (Eng.) Quart. Rev.* **14**, 21–25.

Brown, G. E., and Kennedy, B. W. (1966). Effect of oxygen concentration on Pythium seed rot of soybean. *Phytopathology* **56**, 407–411.

Buller, A. H. R. (1931). "Researches on Fungi," Vol. IV, 329 pp. Longmans, Green, New York.

Burleigh, J. R., Romig, R. W., and Roelfs, A. P. (1969). Characterization of wheat rust epidemics by numbers of urediospores. *Phytopathology* **59**, 1229–1237.

Burleigh, J. R., Eversmeyer, M. G., and Roelfs, A. P. (1972a). Development of linear equations for predicting wheat leaf rust. *Phytopathology* **62**, 947–953.

Burleigh, J. R., Roelfs, A. P., and Eversmeyer, M. G. (1972b). Estimating damage to wheat caused by *Puccinia recondita tritici. Phytopathology* **62**, 944–946.

Caldwell, J. (1933). The physiology of virus diseases of plants. IV. The nature of the virus agent of aucuba or yellow mosaic of tomato. *Ann. Appl. Biol.* **20**, 100–116.

Cammack, R. H. (1961). *Puccinia polysora*: A review of some factors affecting the epiphytotic in West Africa. *Rep. Commonw. Mycol. Conf., 6th, 1960,* pp. 134–138.

Carsner, E., and Lackey, C. F. (1929). Mass action in relation to infection with special reference to curly top of sugar beets. *Phytopathology* **19**, 1137.

Carter, M. V. (1972). How can we improve dose-response studies of pathogens? *Aust. Plant Path. Soc. Newsletter* **1**, 2.

Čech, M., Králik, O., and Blattný, C. (1961). Rod-shaped particles associated with virosis of spruce. *Phytopathology* **51**, 183–185.

Chester, K. S. (1934). Specific quantitative neutralization of the viruses of tobacco mosaic, tobacco ring spot, and cucumber mosaic by immune sera. *Phytopathology* **24**, 1180–1202.

Chester, K. S. (1943). The decisive influence of late winter weather on wheat leaf rust epiphytotics. *Plant Dis. Rep. Suppl.* **143**, 133–144.

Chester, K. S. (1946). "The Nature and Prevention of the Cereal Rusts as Exemplified in the Leaf Rust of Wheat," 269 pp. Chronica Botanica, Waltham, Massachusetts.

Christensen, J. J., and Graham, T. W. (1934). Physiologic specialization and variation in *Helminthosporium gramineum* Rab. *Minn. Univ. Agr. Exp. Sta. Bull.* No. 95, 40 pp.

Clauss, E. (1961). Die phenolischen Inhaltsstoffe der Samenschalen von *Pisum sativum* L. und ihre Bedeutung für Resistenz gegen die Erreger der Fusskrankheit. *Naturwissenschaften* **48**, 106.

Clayton, E. E., and Gaines, J. G. (1945). Temperature in relation to development and control of blue mold (*Peronospora tabacina*) of tobacco. *J. Agr. Res.* **71**, 171–182.

Cohen, M., and Yarwood, C. E. (1952). Temperature response of fungi as a straight line transformation. *Plant Physiolol.* **27**, 634–638.

Colhoun, J. (1961). Spore load, light intensity and plant nutrition as factors influencing the incidence of club root of brassicae. *Trans. Brit. Mycol. Soc.* **44**, 593–600.

Colhoun, J. (1973). Effects of environmental factors on plant disease. *Annu. Rev. Phytopathol.* **11**, 343–364.

Cooper, P. D. (1968). A genetic map of poliovirus temperature-sensitive mutants. *Virology* **35**, 584–596.

Corsten, L. C. A. (1964). Een kwantitatieve beschrijving van de ontwikkeling van een schimmelpopulatie. *Meded. Landbouwhogesch. Wageningen* **64–15**, 1–7.

Crosier, W. (1934). Studies in the biology of *Phytophthora infestans* (Mont.) de Bary. *N.Y. Agr. Exp. Sta. Ithaca Mem.* No. **155**, 1–40.

Crosse, J. E., Goodman, R. N., and Shaffer, W. H. (1972). Leaf damage as a predisposing factor in the infection of apple shoots by *Erwinia amylovora*. *Phytopathology* **62**, 176–182.

Crowe, T. J. (1963). Possible insect vectors of the uredospores of *Hemileia vastatrix* in Kenya. *Trans. Brit. Mycol. Soc.* **46**, 24–26.

Cruickshank, I. A. M., and Perrin, D. R. (1968). The isolation and partial characterization of monicolin A, a polypeptide with phaseollin-inducing activity from *Monilinia fructicola*. *Life Sc.* **7**, 449–458.

Cummins, G. B. (1941). Identity and distribution of the three rusts of corn. *Phytopathology* **31**, 856–857.

Davison, A. D., and Vaughan, E. K. (1964). Effect of urediospore concentration on determination of races of *Uromyces phaseoli* var. *phaseoli. Phytopathology* **54**, 336–338.

Dickson, J. G. (1923). Influence of soil temperature and moisture on the development of the seedling blight of wheat and corn caused by *Gibberella saubinetii. J. Agr. Res.* **23**, 837–869.

Dickson, J. G., Eckerson, S. H., and Link, K. P. (1923). The nature of resistance to seedling blight of cereals. *Proc. Nat. Acad. Sci. U.S.* **9**, 434–439.

Diener, T. O. (1971). Potato spindle tuber "virus". IV. A replicating, low molecular weight RNA. *Virology* **45**, 411–428.

Diener, T. O. (1973). Virus terminology and the viroid: A rebuttal. *Phytopathology* **63**, 1328–1329.

Dimond, A. E. (1941). Measuring inoculum potential and coverage index of sprays. *Phytopathology* **31**, 7.

Dimond, A. E., and Horsfall, J. G. (1960). Prologue. Inoculum and the diseased population. *In* "Plant Pathology" (J. G. Horsfall and A. E. Dimond, eds.), Vol. 3, pp. 1–22. Academic Press, New York.

Dirks, V. A., and Romig, R. W. (1970). Linear models applied to variation in numbers of cereal rust urediospores. *Phytopathology* **60**, 246–251.

Domsch, K. H. (1953). Über den Einflusz photoperiodischer Behandlung auf die Befallsintensität beim Gerstenmehltau. *Arch. Mikrobiol.* **19**, 287–318.

Doubly, J. A., Flor, H. H., and Clagett, C. O. (1960). Relation of antigens of *Melampsora lini* and *Linum usitatissimum* to resistance and susceptibility. *Science* **131**, 229.

Druett, H. A. (1952). Bacterial invasion. *Nature (London)* **170**, 288.

English, P. D., Maglothin, A., Keegstra, K., and Albersheim, P. (1972). A cell wall degrading endopolygalacturonase secreted by *Colletotrichum lindemuthianum*. *Plant Physiol.* **49**, 293–297.

Ercolani, G. L. (1967). Bacterial canker of tomato. II. Interpretation of the aetiology of the quantal response of tomato to *Corynebacterium michiganense* (E.F.Sm.) Jens. by the hypothesis of independent action. *Phytopathol. Mediter.* **6**, 30–40.

Ercolani, G. L. (1973). Two hypotheses on the aetiology of response of plants to phytopathogenic bacteria. *J. Gen. Microbiol.* **75**, 83–95.

Eversmeyer, M. G., and Burleigh, J. R. (1970). A method of predicting epidemic development of wheat leaf rust. *Phytopathology* **60**, 805–811.

Eversmeyer, M. G., Burleigh, J. R., and Roelfs, A. P. (1973). Equations for predicting wheat stem rust development. *Phytopathology* **63**, 348–351.

Favret, E. A. (1971). Basic concepts on induced mutagenesis for disease reaction. *In* "Mutation Breeding for Disease Resistance," pp. 55–65. IAEA, Vienna.

Fehrmann, H. (1963). Untersuchungen zur Pathogenese der durch *Phytophthora infestans* hervorgerufenen Braunfäule der Kartoffelknolle. *Phytopathol. Z.* **46**, 371–408.

Flor, H. H. (1971). Current status of the gene-for-gene concept. *Annu. Rev. Phytopathol.* **9**, 275–296.

Ford, R. E., and Tosic, M (1972). New hosts of maize dwarf mosaic virus and sugarcane mosaic virus and a comparative host range study of viruses affecting corn. *Phytopathol. Z* **75**, 315–348.

Fracker, S. B. (1936). Progressive intensification of uncontrolled plant-disease outbreaks. *J. Econ. Entomol.* **29**, 923–940.

Frandsen, N. O. (1956). Rasse 4 von *Phytophthora infestans* in Deutschland. *Phytopathol. Z.* **26**, 124–130.

Fulton, R. W. (1962). The effect of dilution on necrotic ringspot virus infectivity and the enhancement of infectivity by noninfective virus. *Virology* **18**, 477–485.

Furumoto, W. A., and Mickey, R. (1970). Mathematical analysis of the interference phenomenon of tobacco mosaic virus: Experimental tests. *Virology* **40**, 322–328.

Garren, K. H. (1961). Control of *Sclerotium rolfsii* through cultural practices. *Phytopathology* **51**, 120–124.

Garren, K. H. (1964). Inoculum potential and differences among peanuts in susceptbility to *Sclerotium rolfsii*. *Phytopathology* **54**, 279–281.

Garrett, S. D. (1970). "Pathogenic Root-infecting Fungi," 294 pp. Cambridge Univ. Press, London and New York.

Gäumann, E. (1946). "Pflanzliche Infektionslehre," 611 pp. Birkhaeuser, Basel.

Gavatt, G. F., and Gill, L. S. (1930). Chestnut blight. U.S. *Dep. Agr. Bull.* No. 1641, 18 pp.

Giddings, N. J. (1946). Mass action as a factor in curly-top-virus of sugar beet. *Phytopathology* **36**, 53–56.

Glynne, M. D. (1925). Infection experiments with wart disease of potatoes *Synchytrium endobioticum* (Schilb.) Perc. *Ann. Appl. Biol.* **12**, 34–60.

Goodman, R. N., Király, Z., and Zaitlin, M. (1967). "The Biochemistry and Physiology of Infectious Plant Disease," 354 pp. Van Nostrand—Reinhold, Princeton, New Jersey.

Gorter, G. J. M. A. (1953). Studies on the spread and control of the streak disease of maize. *Dep. Agr. S. Afr. Sci. Bull.* No. 341, 1–20.

Goto, M. (1972). The significance of the vegetation for the survival of plant pathogenic bacteria. *In* "Proceedings of the Third International Conference on Plant Pathogenic Bacteria, 1971" (Maas Geesteranus, ed.), pp. 39–53. Center for Agricultural Publishing and Documentation, Wageningen.

Graham, K. M. (1955). Distribution of physiological races of *Phytophthora infestans* (Mont.) de Bary in Canada. *Amer. Potato J.* **32**, 277–288.

Grainger, J. (1956). Host nutrition and attack by fungal parasites. *Phytopathology* **46**, 445–456.

Grainger, J. (1957). Blight—the potato versus *Phytophthora infestans*. *Agr. Rev.* **3**, 10–26.

Grainger, J. (1959). Effects of diseases on crop plants. *Outlook Agr.* **2**, 114–121.

Green, G. J. (1966). Selfing studies with races 10 and 11 of wheat stem rust. *Can. J. Bot.* **44**, 1255–1260.

Green, G. J. (1971). Physiologic races of wheat stem rust in Canada from 1919 to 1969. *Can. J. Bot.* **49**, 1575–1588.

Gregory, P. H. (1945). The dispersal of air-borne spores. *Trans. Brit. Mycol. Soc.* **28**, 26–72.

Griffin, D. M. (1963a). Soil moisture and the ecology of soil fungi. *Biol. Rev. Cambr. Phil. Soc.* **38**, 141–166.

Griffin, D. M. (1963b). Soil physical factors and the ecology of fungi. II. Behavior of *Pythium ultimum* at small soil water suctions. *Trans. Brit. Mycol. Soc.* **46**, 368–372.

Harrison, B. D., and Woods, R. D. (1966). Serotypes and particle dimensions of tobacco rattle virus with different lengths. *Virology* **28**, 610–620.

Hawker, L. E. (1957). "The Physiology of Reproduction in Fungi," 128 pp. Cambridge Univ. Press, London and New York.

Haymaker, H. H. (1928). Pathogenicity of two strains of the tomato-wilt fungus, *Fusarium lycopersici* Sacc. *J. Agr. Res.* **36**, 675–695.

Heald, F. D. (1921). The relation of spore load to the percent of stinking smut in the crop. *Phytopathology* **11**, 269–278.

Heald, F. D., and Studhalter, R. A. (1914). Birds as carriers of the chestnut blight fungus. *J. Agr. Res.* **2**, 405–422.

Henry, B. W., Moses, C. S., Richards, A., and Riker, A. J. (1944). Oak wilt: Its significance, symptoms and cause. *Phytopathology* **34**, 636–647.

Hepting, G. H. (1971). Diseases of Forest and Shade Trees of the United States, *U.S. Dept. Agr. Forest Ser. Agr. Handb.* No. 386, 658 pp. U.S. GPO, Washington, D.C.

Hildebrand, D. C., and Riddle, B. (1971). Influence of environmental conditions on reactions induced by infiltration of bacteria into plant leaves. *Hilgardia* **41**, 33–43.

Hildebrand, E. M. (1937). Infectivity of the fire-blight organism. *Phytopathology* **27**, 850–852.

Hildebrand, E. M. (1942). A micrurgical study of crown gall infection in tomato. *J. Agr. Res.* **65**, 45–59.

Hirst, J. M. (1958). New methods for studying plant epidemics. *Outlook Agr.* **2**, 16–26.

Hirst, J. M., and Stedman, O. J. (1960). The epidemiology of *Phytophthora infestans*. I. Climate, ecoclimate and the phenology of disease outbreak. *Ann. Appl. Biol.* **48**, 471–488.

Hogen-Esch, J. A., and Zingstra, H. (1969). "Geniteurslijst voor Aartappelrassen," 133 pp. Commissie ter bevordering van het kweken en het ondersoek van nieuwe aartappelrassen, Wageningen, Netherlands.

Hoggan, I. A. (1933). Some factors involved in aphid transmission of the cucumber-mosaic virus to tobacco. *J. Agr. Res.* **47**, 689–704.

Holmes, F. O. (1929). Local lesions in tobacco mosaic. *Bot. Gaz.* **87**, 39–55.

Hooker, A. L., and le Roux, P. M. (1957). Sources of protoplasmic resistance to *Puccinia sorghi* in corn. *Phytopathology* **47**, 187–191.

Horsfall, J. G. (1932). Dusting tomato seed with copper sulfate monohydrate for combating damping-off. *N.Y. State Agr. Exp. Sta. Tech. Bull.* No. 198, 1–34.

Horsfall, J. G. (1938). Combating damping-off. *N.Y. State Agr. Exp. Sta. Tech. Bull.* No. 683, 1–45.

Horsfall, J. G. (1945). "Fungicides and their Action," 240 pp. Chronica Botanica, Waltham, Massachusetts.

Horsfall, J. G., and Dimond, A. E. (1957). Interactions of tissue sugar, growth substances and disease susceptibility. *Z. Pflanzenkr. Pflanzenschutz* **64**, 631–637.

Horsfall, J. G., and Dimond, A. E. (1963). A perspective on inoculum potential. *J. Indian Bot. Soc.* **42A**, 46–57.

Hulett, H. R., and Loring, H. S. (1965). Effect of particle length distribution on infectivity of tobacco mosaic virus. *Virology* **25**, 418–430.

Huttinga, H. (1972). Interaction between long and short particles of tobacco rattle virus. *Meded.* 610 *Inst. Plantenziektenkundig Onderzoek Wageningen*, pp. 80.

Hyre, R. A. (1964). High temperature following infection checks downy mildew of lima beans. *Phytopathology* **54**, 181–184.

Jacobson, M. F., and Baltimore, D. (1968). Polypeptide cleavages in the formation of poliovirus proteins. *Proc. Nat. Acad. Sci. U.S.* **61**, 77–84.

James, W. C., Shih, C. S., Hodgson, W. A., and Callbeck, L. C. (1972). The quantitative relationship between late blight of potato and loss of tuber yield. *Phytopathology* **62**, 92–96.

Jones, T. M., Anderson, A. J., and Albersheim, P. (1972). Host-pathogen interactions. IV. Studies on the polysaccharide-degrading enzymes secreted by *Fusarium oxysporum* f.sp. *lycopersici*. *Physiolol. Plant Pathol.* **2**, 153–166.

Kao, K. N., and Knott, D. R. (1969). The inheritance of pathogenicity to races 111 and 29 of wheat stem rust. *Can. J. Genet. Cytol.* **11**, 266–274.

Kassanis, B., and Nixon, H. L. (1961). Activation of one tobacco necrosis virus by another. *J. Gen. Microbiol.* **25**, 459–471.

Katsuya, K., and Green, G. J. (1967). Reproductive potentials of races 15*B* and 56 of wheat stem rust. *Can. J. Bot.* **45**, 1077–1091.

Keane, P. J. Kerr, A., and New, P. B. (1970). Crown gall of stone fruit. II. Identification and nomenclature of *Agrobacterium* isolates. *Aust. J. Biol. Sci.* **23**, 585–595.

Kelman, A., and Sequeira, L. (1965). Root to root spread of *Pseudomonas solanacearum*. *Phytopathology* **55**, 304–309.

Kerr, A., and Rodrigo, W. R. F. (1967). Epidemiology of tea blister blight (*Exobasidium vexans*). III. Spore deposition and disease production. *Trans. Brit. Mycol. Soc.* **50**, 49–55.

Kerr, A., and Shanmuganathan, N. (1966). Epidemiology of tea blister blight (*Exobasidium vexans*). I. Sporulation. *Trans. Brit. Mycol. Soc.* **49**, 139–145.

Khan, I. D. (1972). Effect of peak vigour and inoculum potential on competitive pathogenic ability and interaction of four cotton root rot pathogens in soil. *Z. Pflanzenkr. Pflanzenschutz* **79**, 714–728.

Killick, R. J., and Malcolmson, J. F. (1973). Inheritance in potatoes of field resistance to late blight [*Phytophthora infestans* (Mont.) de Bary]. *Physiolol. Plant Pathol.* **3**, 121–131.

Király, Z., Barna, B., and Érsek, T. (1972). Hypersensitivity as a consequence, not the cause, of plant resistance to infection. *Nature* (*London*) **239**, 456–457.

Kiyosawa, S. (1971). Genetic approach to the biochemical nature of plant disease resistance. *Jap. Agr. Res. Quart.* **6**, 73–80.

Klebs, G (1900). Zur Physiologie der Fortpflanzung einiger Pilze, *Jahrb. Wiss. Bot.* **35**, 80–203.

Kleczkowski, A. (1950). Interpreting relationships between the concentrations of plant viruses and numbers of local lesions. *J. Gen. Microbiol.* **4**, 53–69.

Klement, Z. (1971). Development of the hypersensitivity reaction induced by plant pathogenic bacteria. *In* "Proceedings of the Third International Conference on Plant Pathogenic Bacteria, 1971" (Maas Geesteranus, ed.), pp. 157–164. Center for Agricultural Publishing and Documentation, Wageningen, Netherlands.

Klement, Z., and Goodman, R. N. (1967a). The hypersensitive reaction to infection by bacterial plant pathogens. *Annu. Rev. Phytopathol.* **5**, 17–44.

Klement, Z., and Goodman, R. N. (1967b). The role of the living bacterial cell and induction time in the hypersensitive reaction of the tobacco plant. *Phytopathology* **57**, 322–323.

Klement, Z., Farkas, G. L., and Lovrekovich, L. (1964). Hypersensitive reaction induced by phytopathogenic bacteria in the tobacco leaf. *Phytopathology* **54**, 474–477.

Knutson, K. W., and Eide, C. J. (1961). Parasitic aggressiveness in *Phytophthora infestans*. *Phytopathology* **51**, 286–290.

Kranz, J. (1968). Eine Analyse von Annuellen Epidemien pilzlicher Parasiten. I. Die Befallskurven und ihre Abhängigkeit von einigen Umweltsfaktoren. *Phytopathol. Z.* **61**, 59–86.

Lampard, J. F., and Carter, G. A. (1973). Chemical investigations on resistance to coffee berry disease in *Coffea arabica*. An antifungal compound in coffee cuticular wax. *Ann. Appl. Biol.* **73**, 31–37.

Lapwood, D. H., and McKee, R. K. (1966). Dose-response relationships for infection of potato leaves by zoospores of *Phytophthora infestans*. *Trans. Brit. Mycol. Soc.* **49**, 679–686.

Last, F. T., and Hamley, R. (1956). A local-lesion technique for measuring the infectivity of conidia of *Botrytis fabae* Sardina. *Ann. Appl. Biol.* **44**, 410–418.

Leach, J. G. (1940). "Insect Transmission of Plant Diseases," 615 pp. McGraw-Hill, New York.

Leach, L. D., and Davey, A. C. (1938). Determining the sclerotial population of *Sclerotium rolfsii* by soil analysis and predicting losses of sugar beets on the basis of these analyses. *J. Agr. Res.* **56**, 619–631.

Limasset, P. (1939). Recherches sur le *Phytophthora infestans* (Mont.) de Bary. *Ann., Epiphyt.* **5**, 21–39.

Lin, K. H. (1939). The number of spores in a pycnidium of *Septoria apii*. *Phytopathology* **29**, 646–647.

Lippincott, J. A., and Heberlein, G. T. (1965a). The infectivity of *Agrobacterium* sp. on Pinto bean leaves. *Amer. J. Bot.* **52**, 633.

Lippincott, J. A., and Heberlein, G. T. (1965b). The quantitative determination of the infectivity of *Agrobacterium tumefaciens*. *Amer. J. Bot.* **52**, 856–863.

Loegering, W. Q., and Powers, H. R. (1962). Inheritance of pathogenicity in a cross of physiologic races 111 and 36 of *Puccinia graminis* f.sp. *tritici*. *Phytopathology* **52**, 547–554.

Logan, C. (1960). Host specificity in two *Xanthomonas* species. *Nature (London)* **188**, 479–480.

Macko, V., Staples, R. C., Renwick, J. A. A., and Pirone, J. (1972). Germination self-inhibitors of rust uredospores. *Physiolol. Plant Pathol.* **2**, 347–455.

Manigault, P., and Beaud, G. (1967). Expression de l'efficacité de la bactérie *Agrobacterium tumefaciens* (Smith *et* Town) Conn. dans l'induction tumorale (*Datura stramonium* L.). *Ann. Inst. Pasteur* **112**, 445–457.

Martin, J. T., Batt, R. F., and Burchill, R. T. (1957). Defense mechanism of plants against fungi: Fungistatic properties of apple leaf wax. *Nature (London)* **180**, 796–797.

Marte, M. (1971). Studies on self-inhibition of *Uromyces fabae* (Pers.) de Bary. *Phytopathol. Z.* **72**, 335–343.

Martinson, C. A. (1963). Inoculum potential relationships of *Rhizoctonia solani* measured with soil microbiological sampling tubes. *Phytopathology* **53**, 634–638.

Massie, L. B., Nelson, R. R., and Tung, G. (1973). Regression equations for predicting sporulation of an isolate of race *T* of *Helminthosporium maydis* on a susceptible male sterile corn hybrid. *Plant Dis. Rep.* **57**, 730–734.

McCallan, S. E. A., and Wellman, R. H. (1943). A greenhouse method for evaluating fungicides by means of tomato foliage diseases. *Contrib. Boyce Thompson Inst.* **13**, 93–134.

McClean, A. P. D. (1947). Some forms of streak virus occurring in maize, sugarcane and wild grasses. *S. Afr. Dep. Agr. Sci. Bull.* No. 265, 39 pp.

McKee, R. K., and Boyd, A. E. W. (1952). Dry-rot disease of the potato. III. A biological method of assessing soil infectivity. *Ann. App. Biol.* **39**, 44–53.

Merrill, W. (1967). The oak wilt epidemics in Pennsylvania and West Virginia: An analysis. *Phytopathology* **57**, 1206–1210.

Meyer, J. A., and Maraite, H. (1971). Multiple infection and symptom expression in vascular wilt disease. *Trans. Brit. Mycol. Soc.* **57**, 371–377.

Mooi, J. C. (1968). Onderzoek inzake de veldresistentie van het loof van de aartappel tegen aantasting door *Phytophthora infestans* (Mont.) de Bary. *Jaarverslag Inst. Plantziektenkundig Onderzoek*, 1968, pp. 125–126.

Mooi, J. C. (1969). The occurrence of races of *Phytophthora infestans* (Mont.) de Bary in the Netherlands in relation to *R*-gene resistance. *Proc. Triennial Conf. Eur. Ass. Potato Res., 4th*, pp. 209–210.

Müller, K. O. (1950). Affinity and reactivity of angiosperms to *Phytophthora infectans*. *Nature (London)* **166**, 392–394.

Müller, K. O. (1953). The nature of the resistance of the potato plant to blight, *Phytophthora infestans*. *J. Nat. Inst. Agr. Bot.* **6**, 346–360.

Müller, K. O., and Börger, H. (1941). Experimentelle Untersuchungen über die Phytophthora-Resistenz der Kartoffel. Zugleich ein Beitrag zum Problem der "erworbenen Resistenz" in Pflanzenreich. *Arb. Biol. Anst. (Reichanst.), Berlin* **23**, 189–231.

Nutman, F. J., and Roberts, F. M. (1963). Studies on the biology of *Hemileia vastatrix* Berk. *et* Br. *Trans. Brit. Mycol. Soc.* **46**, 27–48.

Nutman, F. J., Roberts, F. M., and Bock, K. R. (1960). Method of uredospore dispersal of the coffee leaf-rust fungus, *Hemileia vastatrix*. *Trans. Brit. Mycol. Soc.* **43**, 509–515.

Ohno, S. (1973). Ancient linkage groups and frozen accidents. *Nature (London)* **244**, 259–262

Oort, A. J. P. (1968). A model of the early stage of epidemics. *Neth. J. Plant Pathol.* **74**, 177–180.

Parris, G. K. (1970). "Basic Plant Pathology," 442 pp. State College, Mississippi.

Paxman, G. J. (1963). Variation in *Phytophthora infestans*. *Eur. Potato J.* **6**, 14–23.

Pérombelon, M. C. M. (1972). A quantitative method of assessing virulence of
    Erwinia carotovora var. *carotovora* and *E. carotovora* var. *atroseptica* and
    susceptibility to rotting of potato tuber tissue. *In* "Proceedings of the Third
    International Conference on Plant Pathogenic Bacteria, 1971" (Maas Gees-
    teranus, ed.), pp. 299–303. Center for Agricultural Publishing and Docu-
    mentation, Wageningen.

Person, C. (1959). Gene-for-gene relationships in host: parasite systems. *Can. J.
    Bot.* **37**, 1101–1130.

Person, C., and Cherewick, W. J. (1964). Infection multiplicity in *Ustilago. Can.
    J. Genet. Cytol.* **6**, 12–18.

Petersen, L. J. (1959). Relations between inoculum density and infection of
    wheat by uredospores of *Puccinia graminis* var. *tritici. Phytopathology* **49**,
    607–614.

Peterson, R. S. and Jewell, F. F. (1968). Status of American stem rusts of pine.
    Annu. Rev. Phytopathol. **6**, 23–40.

Peto, S. (1953). A dose-response equation for the invasion of micro-organisms.
    Biometrics **9**, 320–335.

Posnette, A. F., and Robertson, N. F. (1950). Virus diseases of cacao in West
    Africa. VI. Vector investigations. *Ann. App. Biol.* **37**, 363–377.

Posnette, A. F., Robertson, N. F., and Todd, J. M. (1950). Virus diseases of
    cacao in West Africa. V. Alternative host plants. *Ann. Appl. Biol.* **37**,
    229–240.

Price, W. C. (1938). Studies on the virus of tobacco necrosis. *Amer. J. Bot.* **25**,
    603–612.

Price, W. C., and Spencer, E. L. (1943). Accuracy of the local lesion method
    for measuring virus activity. II. Tobacco-necrosis, alfalfa mosaic and tobacco
    ringspot viruses. *Amer. J. Bot.* **30**, 340–346.

Rapilly, F. (1968). Quelques remarques sur la morphologie des urédospores de
    Puccinia striiformis f. sp. *tritici. Bull. Grim. Soc. Mycol. France* **84**,
    493–496.

Rapilly, F., and Fournet J. (1968). Observations sur la dissémination du *Puccinia
    striiformis* en fonction de l'humidité relative, relation avec la structure mor-
    phologique des urédospores. Cereal Rust Conf, 1968, Oeiras.

Rapilly, F., Fournet, J., and Skajennikoff, M. (1970). Études sur l'épidémiologie
    et la biologie de la rouille jaune du blé *Puccinia striiformis* Westend. *Ann.
    Phytopathol.* **2**, 5–31.

Raymer, W. B., and Diener, T. O. (1969). Potato spindle tuber virus: A plant
    virus with properties of a free nucleic acid. I. Assay, extraction and concen-
    tration. *Virology* **37**, 343–350.

Roelfs, A. P., and McVey, D. V. (1972). Wheat stem rust races in the Yaqui
    valley of Mexico during 1972. *Plant Dis. Rep.* **56**, 1038–1039.

Rotem, J., Cohen, Y., and Putter, J. (1971). Relativity of limiting and optimum
    inoculum loads, wetting durations, and temperatures for infection by *Phytoph-
    thora infestans. Phytopathology* **61**, 275–278.

Rowell, J. B., and Olien, C. R. (1957). Controlled inoculations of wheat seed-
    lings with uredospores of *Puccinia graminis* var. *tritici. Phytopathology* **47**,
    650–655.

Royle, D. J. (1973). Quantitative relationships between infection by the hop downy mildew pathogen, *Pseudoperonspora humuli,* and weather and inoculum factors. *Ann. Appl. Biol.* **73**, 19–30.

Sadasivan, T. S., and Subramanian, C. V. (1969). Interaction of pathogen, soil, other microorganism in the soil, and host. *In* "Plant Pathology," Horsfall and A. E. Dimond, eds.), Vol. 2, pp. 273–313. Academic Press, New York.

Samuel, G., and Bald, J. G. (1933). On the use of the primary lesions in quantitative work with two plant viruses. *Ann. Appl. Biol.* **20**, 70–99.

Sänger, H. L. (1968a). Characteristics of tobacco rattle virus. I. Evidence that its two particles are functionally defective and mutually complementing. *Mol. Gen. Genet.* **101**, 346–367.

Sänger, H. L. (1968b). Defective plant viruses. "Mol. Gen. Proc. Wiss. Konf. Gesellsch. Deutsch. Naturforsch Ärzte, Berlin, *4th, 1967"* pp. 300–336. Springer-Verlag, Berlin.

Scheffer, T. C., and Cowling, E. B. (1966). Natural resistance of wood to microbial deterioration. *Annu. Rev. Phytopathol.* **4**, 147–170.

Schein, R. D. (1964). Design, performance and use of a quantitative inoculator *Phytopathology* **54**, 509–513.

Schnathorst, W. C., and De Vay, J. E. (1963). Common antigens in *Xanthomonas malvacearum* and *Gossypium hirsutum* and their possible relationship to host specificity and disease resistance. *Phytopathology* **53**, 1142.

Schneider, I. R. (1971). Characteristics of a satellite-like virus of tobacco ringspot virus. *Virology* **45**, 108–122.

Schrödter, H. (1960). Dispersal by air and water—the flight and landing. *In* "Plant Pathology" (J. C. Horsfall and A. E. Dimond, eds.), Vol. 3, pp. 169–227. Academic Press, New York.

Schrödter, H. (1965). Methodisches zur Bearbeitung phytometeoropathologischer Untersuchungen, dargestellt am Beispiel der Temperaturrelation. *Phytopathol. Z.* **53**, 154–166.

Schrödter, H., and Ullrich, J. (1965). Untersuchungen zur Biometeorologie und Epidemiologie vor *Phytophthora infestans* (Mont.) de By. auf mathematisch-statisticher Grundlage. *Phytopathol. Z.* **54**, 87–103.

Segall, R. H., and Newhall, A. G. (1960). Onion blast or leaf spotting caused by a species of *Botrytis. Phytopathology* **50**, 76–82.

Semancik, J. S., Morris, T. J., and Weathers, L. G. (1973). Structure and conformation of low molecular weight RNA from exocortis disease. *Virology* **53**, 448–456.

Semancik, J. S., and Weathers, L. G. (1972). Pathogenic 10 S RNA from exocortis disease recovered from tomato bunchy-top plants similar to potato spindle tuber virus infection. *Virology* **49**, 622–625.

Semeniuk, G. (1965). Comment on a paper by Dimond and Horsfall at the International Symposium on Factors Determining the Behavior of Plant Pathogens in the Soil, Berkeley, Calif., 1963. *In* "Ecology of Soil borne Plant Pathogens" (K. F. Baker and W. C. Snyder, eds.), p. 415. Murray, London.

Shaner, G. E., Peart, R. M., Newman, J. E., Stirm, W. L., and Loewer, O. L. (1972). A plant disease display model: an evaluation of the computer simulator EPIMAY for southern corn leaf blight in Indiana *Purdue Univ. Agr. Exp. Sta. Publ.* No. RB–890, 15 pp.

Shearer, B. L., and Zadoks, J. C. (1972). The latent period of *Septoria nodorum* in wheat. I. The effect of temperature and moisture under controlled conditions. *Neth. J. Plant Pathol.* **78**, 231–241.

Simmonds, N. W., and Malcolmson, J. F. (1967). Resistance to late blight in *Andigena* potatoes. *Eur. Potato J.* **10**, 161–166.

Singh, R. P (1971). A local lesion host for potato spindle tuber virus. *Phytopathology* **61**, 1034–1035.

Sogo, J. M., Koller, T., and Diener, T. O. (1973). Potato spindle tuber viroid.X. Visualization and size determination by electron microscopy. *Virology* **55**, 70–80.

Stanbridge, B., and Gay, J. L. (1969). An electron microscope examination of surfaces of the uredospores of four races of *Puccinia striiformis*. *Trans. Brit. Mycol. Soc.* **53**, 149–153.

Stanton, W. R., and Cammack, R. H. (1953). Resistance to the maize rust *Puccinia polysora* Underw. *Nature (London)* **172**, 505–506.

Stewart, D. M., Romig, R. W., and Rothman, P. G. (1970). Distribution and prevalence of physiologic races of *Puccinia graminis* in the United States in 1968. *Plant Dis. Rep.* **54**, 256–260.

Stholosuta, P., Bailey, J. A., Severin, V., and Deverall, B. J. (1971). Effect of bacterial inoculation of bean and pea leaves on the accumulation of phaseollin and pisatin. *Physiol. Plant Pathol.* **1**, 177–183.

Storey, H. H. (1938). Investigations on the mechanism of the transmission of plant viruses by insect vectors. II. The part played by puncture in transmissions. *Proc. Royal Soc. Ser. B.* **125**, 455–477.

Storey, H. H., Howland, A. K., Hemingway, J. S., Jameson, J. D., Baldwin, B. J. T., Thorpe, H. C., and Dixon, G. E. (1958). East African work on breeding maize resistant to tropical American rust *Puccinia polysora*. *Emp. J. Exp. Agr.* **26**, 1–17.

Teakle, D. S. (1973). Use of the local lesion method to study the effect of celite and inhibitors on virus infection of roots. *Phytopathol. Z.* **77**, 209–215.

Thresh, J. M. (1958). The spread of virus disease in cacao. *West Afr. Cocoa Res. Inst. Tech. Bull.* No. 5, 36 pp.

Thurston, H. D. (1961). The relative survival ability of races of *Phytophthora infestans* in mixtures. *Phytopathology* **51**, 748–755.

Thurston, H. D., and Eide, C. J. (1952). The appearance and survival of new races of *Phytophthora infestans*. *Phytopathology* **42**, 481–482.

Thurston, H. D., and Eide, C. J. (1953). The survival of races of *Phytophthora infestans* in a mixture. *Phytopathology* **43**, 486.

Thyr, B. D. (1968). Bacterial canker of tomato: inoculum level needed for infection. *Plant Dis. Repr.* **52**, 741–743.

Toxopeus, H. J. (1956). Reflections on the origin of new physiologic races in *Phytophthora infestans* and the breeding of resistance in potatoes. *Euphytica* **5**, 221–237.

Trevan, J. W. (1927). The error of determination of toxicity. *Proc. Royal Soc. Ser. B* **101**, 483–514.

Van Arsdel, E. P. (1965). Micrometeorology and plant disease epidemiology. *Phytopathology* **55**, 945–950.

Van der Plank, J. E. (1960). Analysis of epidemics. *In* "Plant Pathology" (J. G. Horsfall and A. E. Dimond, eds.), Vol. 3, pp. 229–287. Academic Press, New York.

Van der Plank, J. E. (1963). "Plant Diseases: Epidemics and Control," 349 pp. Academic Press, New York.

Van der Plank, J. E. (1965). Dynamics of epidemics of plant disease. *Science* **147**, 120–124.

Van der Plank, J. E. (1967a). Epidemiology of fungicidal action. *In* "Fungicides" (D. C. Torgeson, ed.), Vol. 1, pp. 63–92. Academic Press, New York.

Van der Plank, J. E. (1967b). Spread of plant pathogens in space and time. *In* "Airborne Microbes" (P. H. Gregory and J. L. Monteith, eds.), 227–246. Cambridge Univ. Press, London and New York.

Van der Plank, J. E. (1968). "Disease Resistance in Plants," 206 pp. Academic Press, New York.

Van der Plank, J. E. (1971). Stability of resistance to *Phytophthora infestans* in cultivars without *R* genes. *Potato Res.* **14**, 263–270.

Van Kammen, A. (1968). The relationship between the components of cowpea mosaic virus. I. Two ribonucleoprotein particles necessary for infectivity of CPMV. *Virology* **34**, 312–318.

Van Vloten-Doting, L., Dingjan-Versteegh, A., and Jaspars, E.M.J. (1970). Three nucleoprotein components of alfalfa mosaic virus necessary for infectivity. *Virology* **4**, 419–430.

Waggoner, P.E. (1952). Distribution of potato late blight around inoculum sources. *Phytopathology* **42**, 323–328.

Waggoner, P. E., and Horsfall, J. G. (1969). EPIDEM: A simulator of plant disease written for a computer. *Conn. Agr. Exp. Sta. Bull.* No. 698, 80 pp.

Waggoner, P. E., Horsfall, J. G., and Lukens, R. J. (1972). EPIMAY: A simulator of southern corn leaf blight. *Conn. Agr. Exp. Sta. Bull.* No. 729, 84 pp.

Walker, J. C., and Link, K. P. (1935). Toxicity of phenolic compounds to certain onion bulb parasites. *Bot. Gaz.* **96**, 468–484.

Wallin, J. R., and Hoyman, W. G. (1958). Influence of post-inoculation air temperature maxima on survival of *Phytophthora infestans* in potato leaves. *Amer. Potato J.* **35**, 769–773.

Warren, R. C., King, J. E., and Colhoun, J. (1973). Reaction of potato plants to *Phytophthora infestans* in relation to their carbohydrate content. *Britt. Mycol. Soc. Trans.* **61**, 95–105.

Wastie, R. L. (1962). Mechanism of action of an infective dose of *Botrytis* spores on bean leaves. *Trans. Brit. Mycol. Soc.* **45**, 465–473.

Watson, M. A. (1936). Factors affecting the amount of infection obtained by aphis transmission of the virus *Hy* III. *Phil. Trans. Royal Soc. London Ser. B* **226**, 457–489.

Weihing, J. L., and O'Keefe, R. B. (1962). Epidemiological potentials of potato varieties in relation to late blight. *Phytopathology* **52**, 1268–1273.

White, J. H. (1919). On the biology of *Fomes applanatus* (Pers.) Wallr. *Trans. Roy. Can. Inst.* **14**, 133–174.

Wilhelm, S. (1950). Vertical distribution of *Verticillium albo-atrum* in soils. *Phytopathology* **40**, 368–376.

Wilhelm, S. (1951). Effect of various soil amendments on the inoculum potential of the Verticillium wilt fungus. *Phytopathology* **41**, 684–690.

Williams, N. D., Gough, F. J., and Rondon, M. R. (1966). Interaction of pathogenicity genes in *Puccinia graminis* f.s.p. *tritici* and reaction genes in *Triticum aestivum* ssp. *vulgare* "Marquis" and "Reliance". *Crop. Sci.* **6**, 245–248.

Wilson, E. M. (1958). Aspartic and glutamic acid as self-inhititors of uredospore germination. *Phytopathology* **48**, 595–600.

Wimalajeewa, D. L. S., and De Vay, J. E. (1971). The occurrence and characterization of a common antigen relationship between *Ustilago maydis* and *Zea mays. Physiol. Plant Pathol.* **1**, 523–535.

Wolfe, M. S. (1972). The genetics of barley mildew. *Rev. Plant Pathol.* **51**, 507–522.

Wood, R. K. S. (1967). "Physiological Plant Pathology," 570 pp. Blackwell Oxford.

Woodbury, W., and Stahmann, M. A. (1970). Role of surface films in the germination of rust uredospores. *Can. J. Bot.* **48**, 499–511.

Yarwood, C. E. (1954). Mechanism of acquired immunity to a plant rust. *Proc. Nat. Acad. Sci. U.S.* **40**, 374–377.

Youden, W. J., Beale, H. P., and Guthrie, J. D. (1935). Relation of virus concentration to the number of lesions produced. *Contrib. Boyce Thompson Inst.* **7**, 37–53.

Zadoks, J. C. (1972). Methodology of epidemiological research. *Annu. Rev. Phytopathol.* **10**, 253–276.

Subject Index

A

Aggressiveness of pathogens, 192
Agrobacterium radiobacter, infection by single cells, 25
Alfalfa mosaic virus, see Synergism, obligate and Vectors transmitting viruses
Alternaria solani
disease/inoculum curve in tomato, 58
epidemics of, 101-102
temperature effect, 83
Antigens, shared by host and parasite in compatible combinations, 182-184
Armillaria mellea, host/temperature interactions during infection, 83

B

Barley mildew, see Erysiphe graminis
Bean (Phaseolus vulgaris), see Colletotrichum lindemuthianum and Uromyces phaseoli
Bean (Vicia faba), see Botrytis cinerea and Botrytis fabae
Botrytis allii, disease/inoculum curve on onion, 58
Botrytis cinerea, disease/inoculum curve on bean (Vicia faba), 58, 64
Botrytis fabae, disease/inoculum curve on bean (Vicia faba), 3, 53, 58, 64, 69

C

Carbohydrates in relation to resistance, 170-171
Ceratocystis fagacearum, see Oak wilt disease
Citrus exocortis virus, 9-10, 36, 126, 184
Citrus tristeza virus, 126
Citrus xyloporosis virus, 126
Coffee, see Hemileia vastatrix
Colletotrichum lindemuthianum
resistance, horizontal, to, in bean, 168-169

resistance, vertical, to, in bean, 167-168
Corn, see Maize
Corynebacterium michiganense
disease/inoculum curves in tomato, 25-26, 55-56
multiple infection in tomato, 12
Cotton, see Fusarium vasinfectum, Rhizoctonia solani, Sclerotium bataticola, and Sclerotium rolfsii
Cowpea mosaic virus, see Synergism, obligate and Vectors transmitting viruses
Cronartium comandrae, epidemics in pine, 120
Cronartium fusiforme, epidemics in pine, 119-120
Cronartium occidentale, epidemics in pine, 120
Cronartium ribicola
continuity of epidemics in pine, 99
in epidemic and endemic disease, 112-113, 121-122
waves of infection in pine, 96

D

Disease/inoculum curves, see also individual pathogens
for antagonistic interaction between spores, 3-4, 10-11, 62-63
artifacts in, 29-31, 59-61
for bacteria, 25 ff.
for competition between spores, 3, 10-11, 62-63
in relation to ED_{50}, 58
for facultative synergism between spores, 4-6, 10-11, 62-63
for independent action without competition or interaction between spores, 2, 10-11, 46-47, 62-63
in relation to the number and quality of susceptible sites, 49-58
for obligate synergism during infection, 22-24
near the origin, 45-47

211

Disease/inoculum, curves (cont'd)
slope of, 61-62, 74
with vector transmission, 40 ff.
for viruses, 29 ff.
Dispersal of pathogens, see also Spread
of disease, and individual pathogens
gradients, 135-138

E

Ecology determining host range of
viruses, see Forest trees, Maize
dwarf mosaic virus, Maize mosaic
virus, Maize streak virus, Sorghum red stripe virus, and Sugarcane mosaic virus
Elm phloem necrosis, 126
Endemic disease, see also individual
pathogens
adaptation of pathogens to, 122-123
in continuum with epidemic disease,
113 ff.
implications of endemicity, 120-122
limit (asymptote) of disease, 114-115
steady state impossible, 117
time relatively unimportant, 112 ff.
Endothia parasitica
constancy of infection rate, 100
destructiveness, 129
dispersal, 138
history in America, 131
Epidemics, principle of continuity, 100
Erwinia amylovora, infection by single
cells, 25-26
Erwinia carotovora var. atroseptica, infection by single cells, 26
Erysiphe graminis, disease/inoculum
curve on barley, 4, 60
Exobasidium vexans, spore production,
79

F

Flax rust, see also Antigens
gene-for-gene relation, 145-147
Foci of disease, see also Spread of
disease, 114, 134 ff.

Fomes applanatus, spore production,
122-123
Forest trees, comparative rarity of
harmful virus infections, 124 ff.
Fusarium caeruleum, disease/inoculum
curves in potato tubers, 53-55
Fusarium oxysporum f. sp. lycopersici,
disease/inoculum curve in tomatoes, 16-17
Fusarium oxysporum f. sp. melonis, see
Multiple infection
Fusarium vasinfectum, disease/inocu-
lum curve in cotton, 60

G

Gene-for-gene relations
absence in horizontal resistance, 144
first (Flor's) hypothesis, 145-147
second hypothesis, 147-148, 151-152,
155-158
Germination of spores
self-inhibition, 3-4, 59-60
synergistic interaction, 4-5, 60-62
water relations, 58-59
Gibberella saubinetii
disease/inoculum curve, 16-18
host/temperature interaction during
infection, 83
Grape powdery mildew, see Uncinula
necator

H

Helminthosporium maydis
dew period and temperature, effect of,
94, 103
epidemics of, 102-103
Hemileia vastatrix
disease/inoculum curve, 18-19
dispersal and spread, 138
Heterogenesis of viruses, evidence
against, 8-10
Hypersensitivity, see also Phytoalexins
association with vertical resistance
and immunity, 181
effect not necessarily macroscopically
visible, 28, 33-34

I

Immunity from disease, *see also* Population immunity
  definition, 143, 181
  as extreme vertical resistance, 181-182
  through forbidden virulence, 164, 190-192
  host range determination, 122
Incubation period of disease, defined, 93, 109
Independent action of spores during infection, tests for, 67-68
Infection, continuous versus discontinuous, 105-107
Infection rate *r*
  definition, 92, 104
  historic (memory) factors, effect on variance, 94-100
  waves of infection, 95 ff., 110
Infection rate *R*
  definition, 91-92, 104
  infection rate *r*, comparison with, 93-94
  infection rate *r*, relation with, 105
  product *iR* with period of infectiousness *i*, 114 ff., 130
Infectious entities, 42-44
Inoculum *see also* Disease/inoculum curves and Inoculum potential
  importance in epidemics, 80, 102-103
Inoculum potential, 85 ff.
Inoculum's potential, 84-85

L

Law of the origin, 20-21
Logarithm of inoculum, to be avoided, 68-70

M

Maize dwarf mosaic virus, host range, 128
Maize mosaic virus, host range, 128
Maize rust, *see Puccinia sorghi*
Maize streak virus
  disease/inoculum curve in vector transmission, 40, 42
  host range, 127-128

Maize tropical rust, *see Puccinia polysora*
Melampsora lini, *see* Flax rust
Metavirus, 9
Moisture, effect of, 58-59, 81-84
Multiple infection, correction for, 12-14
Multiple regression analysis, *see also* Infection rate *r*, 77-80

N

Numerical threshold of infection, *see also* Synergism, obligate
  confusion with dilution end point, 19-20
  undemonstrated, 6-8, 14 ff.

O

Oak wilt disease
  epidemics, antiepidemics, and endemicity, 117-119
  as systemic wilt disease, 129
Orange, *see* Citrus
Oxygen in relation to infection, 76-77

P

Parameters in disease/inoculum relations, *see also* Multiple regression analysis, 70-73
Pathogenicity, *see* Aggressiveness of pathogens, and Virulence of pathogens
Penicillium digitatum, infection in relation to the quality and quantity of wounds in oranges, 48-51
Peridermium filamentosum, epidemics in pine, 120
Peridermium harknessii, epidemics in pine, 120
Period of infectiousness *i*, *see also* Infection rate *R*
  definition, 92
  in discontinuous infection, 107
  in disease increase, equations for, 104-105, 114
  in endemic disease, 114 ff.

Period of latency *p*
  continuity of epidemics, role in, 100
  in continuous and discontinuous infection, 105-107
  definition, 92, 107 ff.
  effect of infection rate *r* on, 108
  importance in explosive epidemics, 122
  unimportance in timeless disease, 114
  wavelength of epidemics, role in, 96
Peronospora tabacina
  effect of temperature and light on sporulation, 84
  epidemics of, 131
Phytoalexins
  association with vertical resistance and immunity, 181
  Müller's theory of postinfection antibiotic activity, 187
  oddity of history, 189
  production following chemical injury, 186
  result of pathogen's death, 186
  Van der Plank's theory of localized, preformed antifungal compounds, 188-189
Phytophthora infestans, see also Carbohydrates
  disease/inoculum curves in potato and tomato, 58
  dispersal of spores, 135-138
  infection, factors governing, 78
  moisture, effect of, 82-84
  resistance, horizontal, in potato, 166-167, 178-179
  resistance, vertical, genes R_1 and R_4 in potato, 148 ff.
  temperature, effect of, 81, 83-84
  variation in epidemic progress, 78
  virulence frequency in populations, 148 ff., 165
Phytophthora phaseoli, effect of temperature, 81
Pine species susceptible to *Cronartium ribicola,* 122
Plasmodiophora brassicae, disease/inoculum curve, 57-58

Poisson curves for "multiple hit" infection
  equations, 24
  rules for confirming, 33-34
Poisson curves for "one hit" infection, 22, 24, 46-47, 49-52, 70-72
Polygalacturonase, 169-170
Population immunity, condition for, 130
Potato, *see Alternaria solani, Fusarium caeruleum, Phytophthora infestans,* and *Synchytrium endobioticum*
Potato spindle tuber virus, 9-10, 36, 184-185
Prune dwarf virus, *see* Synergism, obligate
Pseudomonas spp., infection by single cells, 26
Pseudomonas solanacearum, facultative synergism during infection, 62
Pseudoperonospora humuli, humidity in relation to infection, 76-79
Puccinia graminis, see also Wheat stem rust
  disease/inoculum curve in wheat, 4-5, 57
  frequency of virulence on Sr_5, Sr_6, and Sr_{9d} in populations, 152-157
  geographical changes in virulence, 161-162
  synergism, facultative, during spore germination, 4-5
Puccinia polysora, resistance in maize to, 177-178, 180
Puccinia recondita, see Wheat leaf rust
Puccinia sorghi, resistance in maize to, 176-177
Puccinia striiformis
  spore clumps of, 6, 43
  temperature in relation to, 83
Pythium ultimum, tolerance of poor soil aeration, 76-77

R

Relative humidity, *see* Moisture
Removal of disease, *see also* Period of infectiousness, 92

Resistance, horizontal, to disease, *see also Colletotrichum lindemuthianum* and *Phytophthora infestans*
commonness of, 170-172
definition, 144, 166
in endemic disease, 121
excess harmful to host, 121, 179-180
molecular theory of enzyme dose, 174
quantity of, 176-179
terminology, 193-194
variation curbed by evolutionary freeze, 175-176
Resistance, vertical, to disease, *see also Colletotrichum lindemuthianum, Phytophthora infestans,* and *Puccinia graminis*
definition, 144
gene diversification, 173-174
in relation to the gene-for-gene hypotheses, 144, 158
through new genes, 172-173
terminology, 193-194
weak genes, 160
Rhizoctonia solani, disease/inoculum curve in cotton, 60
Ribonucleic acid, infectious, 9-10, 184-185
Robinia witches' broom, 124, 126
Roots, disease/inoculum curve not unique, 63-66

S

Satellitism in viruses, 36-37
Sclerotium bataticola, disease/inoculum curve in cotton, 60
Sclerotium rolfsii
disease/inoculum curve in beet, 53, and cotton, 60
food bases in infection of peanuts, 77
Selection, directional, 164
Selection, stabilizing, 155, 164
Septoria lycopersici, disease/inoculum curve in tomato, 58
Sorghum red stripe virus, natural host range, 128
Sour cherry ringspot virus, *see* Synergism, obligate

Spread of disease
defined, 131, 133
feature of epidemic disease, 132
foci of disease in relation to, 134-137
infection rate in relation to, 132-133, 138 ff.
as migration and colonization by pathogen, 137
Sugar beet curly-top virus, dose of, in relation to transmission by vectors, 40
Sugarcane mosaic virus, host range, 128
Sugarcane streak virus, *see* Maize streak virus
Susceptibility to infection, two-path concept of, 48-58
Synchytrium endobioticum, disease/inoculum curve in potato, 6-8, 14-16
Synergism, facultative, during infection
in *Pseudomonas solancearum,* 62
in *Puccinia graminis,* 4,5, 60-62
in *Puccinia striiformis,* 6, 43
obligate, during infection, *see also* Numerical threshold of infection
absence during infection by bacteria, 25-28
absence during infection by fungi, 6-8, 14-19
absence during infection by tobacco mosaic virus, 30-32
consequence of qualitative deficiency, 44-45
as genetic supplementation, 35
presence during infection by alfalfa mosaic, cowpea mosaic, prune dwarf, sour cherry ringspot, and tobacco rattle viruses, 32 ff.

T

Temperature
effect of, 80-81
interaction with other factors, 83-84, 94

Theobroma viruses
  in Bombacaceae and Sterculiaceae, 126-127
  disease/inoculum curve with vector transmission, 40-41
  mechanism of tolerance to, 126
Time
  as dimension, 88 ff., 131 ff.
  timeless disease, equation for, 114-115
Tobacco mosaic virus, disease/inoculum curves, 29-32
Tobacco necrosis virus, disease/inoculum curves, 32, 42
Tobacco rattle virus, *see* Synergism, obligate and Vectors transmitting viruses
Tomato, *see Alternaria solani, Corynebacterium michiganense, Phytophthora infestans, Septoria lycopersici,* and Tomato bunchy top virus
Tomato bunchy top virus, 9-10, 184
Tranzschelia discolor
  disease/inoculum curve on peach, 19
  spore germination, 59

U

Uncinula necator, 179-180
Units of dispersal, 6, 43
Uromyces phaseoli, disease/inoculum curves on bean, 1-4, 52-53, 57, 59-60
Ustilago avenae, see Multiple infection
Ustilago maydis, see Antigens
Ustilago kolleri, see Multiple infection

V

Vectors transmitting viruses, *see also* Disease/inoculum curves in relation to obligate synergism between particles, 39, 42
Vertifolia effect, the loss of horizontal resistance while breeding new varieties for vertical resistance, 179
Viroid, *see* Citrus exocortis virus, Potato spindle tuber virus, and Tomato bunchy top virus
Virulence of pathogens, *see also* Gene-for-gene relations
  by adaptation, 163
  definition, 192
  forbidden, 164
  inheritance, 183
  as missense mutation, 157-158
  preexisting, 162
  restricted, 163
Virus *Hy* III, disease/inoculum curve in vector transmission, 40-41

W

Wheat leaf rust
  prediction of epidemics, 78-79
  variables affecting, 97-98
Wheat stem rust, *see also Puccinia graminis*
  prediction of epidemics, 78-79
  temperature in relation to, 83
  waves of infection, 97

X

Xanthomonas citri, infection by single cells, 25

A
B
C
D
E
F
G
H
I
J